THINKING VISUALLY

THINKING VISUALLY

STEPHEN K. REED

Routledge
Taylor & Francis Group

LONDON AND NEW YORK

First published in 2010 by Psychology Press

Published 2016 by Routledge
2 Park Square, Milton Park, Abingdon, Oxfordshire OX14 4RN
711 Third Avenue, New York, NY 10017

First issued in paperback 2016

Routledge is an imprint of the Taylor and Francis Group, an informa business

© 2010 by Taylor and Francis Group, LLC

International Standard Book Number: 978-1-138-99030-2 (Paperback)
International Standard Book Number: 978-0-8058-6067-2 (Hardback)

All rights reserved. No part of this book may be reprinted or reproduced or utilised in any form or by any electronic, mechanical, or other means, now known or hereafter invented, including photocopying and recording, or in any information storage or retrieval system, without permission in writing from the publishers.

Notice:
Product or corporate names may be trademarks or registered trademarks, and are used only for identification and explanation without intent to infringe.

For permission to photocopy or use material electronically from this work, please access www.copyright.com (http://www.copyright.com/) or contact the Copyright Clearance Center, Inc. (CCC), 222 Rosewood Drive, Danvers, MA 01923, 978-750-8400. CCC is a not-for-profit organization that provides licenses and registration for a variety of users. For organizations that have been granted a photocopy license by the CCC, a separate system of payment has been arranged.

**Visit the Taylor & Francis Web site at
http://www.taylorandfrancis.com**

In Memory of
Tom Trabasso
Herb Simon
Jim Kaput

I dedicate this book to the memory of three men who had important influences on my career. Tom Trabasso was one of the first to welcome me when I began my graduate study at UCLA. He left for Princeton before I completed my graduate study but, before leaving, gave me a copy of a 1968 article by N. S. Sutherland titled "Outlines of a Theory of Visual Pattern Recognition in Animals and Man." That article led to my spending a post-doc with Sutherland at the University of Sussex. While at Sussex, I was invited to apply for a faculty position at Case Western Reserve. I later discovered that Tom had recommended me for this position. Shortly after joining its faculty, I published my dissertation (Pattern Recognition and Categorization) in *Cognitive Psychology* with much help from the action editor. You can probably guess his name.

Although Tom helped me launch my career, Herb Simon inspired me to emphasize problem solving in my later research. In 1975, I was invited to spend a year at Carnegie Mellon University, which gave me the opportunity to work with Herb on a project. I had collected data showing that a subgoal helped students solve the missionaries and cannibals problem and Herb suggested that we build a detailed computer simulation model to determine why the subgoal was successful (discussed in Chapter 4). My contribution was mostly to sit back and watch a genius at work. I also quickly learned that Herb was as supportive of students and colleagues as he was brilliant.

Jim Kaput had the most recent influence on my career. Jim believed that algebra was a major hurdle in the path to many careers and that all students could succeed with the right kind of support. He and Jeremy Roschelle developed the SimCalc software to make this happen. I hope our Animation Tutor project at San Diego State will contribute to Jim's vision. Chapter 14 discusses these projects.

Contents

Preface xiii
Acknowledgments xv

Chapter 1   Images Versus Words 1
                 Thinking With Symbols 3
                 Propositions 5
                 Perceptual Symbols 6
                 Organization of the Book 10
                 Summary 11

Chapter 2   Images Before Words 13
                 Visual Memory 13
                 From Memory to Concepts 14
                 Small Quantities 17
                 Space, Quantity, and Time 18
                 Numerical Reasoning With Language 22
                 Summary 23

Chapter 3   Estimation 25
                 The Mental Number Line 26
                 Spatial Skills 27
                 Measurement 28
                 Mixture Problems 30
                 Interpolation 33
                 Summary 34

Chapter 4   Spatial Metaphors 35
                 Applications to Cognitive Psychology 36
                 Applications to Mathematics 37
                 Grounding Metaphors 40
                 Searching Search Spaces 42
                 Linking Metaphors 45
                 Summary 46

Chapter 5   Producing Images 47
                 Remembering Visual Images 47
                 Advantages of Images 49

	Prototypical Images.. 51
	Limitations of Images 53
	Reality Monitoring 55
	Breakdown of Reality Monitoring 56
	Summary 57
Chapter 6	Manipulating Images.. 59
	Scanning Images 59
	Causal Reasoning 64
	Rearranging Objects 66
	Designing With Images 68
	The Geneplore Model.. 71
	Summary 73
Chapter 7	Viewing Pictures 75
	Evocative Pictures 75
	Decorative Pictures 76
	Static Pictures.. 79
	Animated Pictures... 82
	Static Versus Animated Pictures............................ 84
	Summary 86
Chapter 8	Producing Diagrams.. 87
	Constructing Representations................................ 88
	Emphasizing Spatial Relations 89
	Representing Space and Time 91
	Matching Diagrams to Problems 92
	Spatial Representations of Meaning...................... 95
	Semantic Networks as Instructional Tools............. 96
	Summary 98
Chapter 9	Comprehending Graphs 99
	Reasoning From Graphs... 99
	Interpreting Events Over Time............................... 102
	The Algebra Sketchbook 104
	Exponential Growth 106
	Modeling Population Growth.................................. 108
	Summary 111

Contents ix

Chapter 10 Words and Pictures..113
  Integration in Working Memory 113
  Cognitive Load Theory 115
  Mayer's Multimedia Theory...117
  Words and Images 119
  Simulated Actions 120
  Inferences 122
  Summary 123

Chapter 11 Vision and Action.. 125
  Memory for Actions 125
  Acting.. 126
  Chess 128
  Reasoning.. 129
  Reflections 131
  Manipulatives.. 132
  Summary 134

Chapter 12 Virtual Reality... 135
  Examples of Virtual Reality.. 135
  Exploring Environments 136
  Ecological Psychology and Multimedia...................... 137
  Situation Awareness 139
  Virtual Military Training...141
  Virtual Biking 143
  Virtual Driving 144
  Summary 144

Chapter 13 Science Instructional Software.. 147
  Physics 148
  Ecological Systems... 150
  Transfer of Principles 153
  Chemistry.. 156
  Scientific Design... 159
  Summary 160

Chapter 14 Mathematics Instructional Software 161
  SimCalc 161
  The Animation Tutor™.. 164
  Object Manipulation... 165
  Simulation 167

ANIMATE 169
Summary 171

Chapter 15 Conclusions 173

Multimedia Learning 174
The Need 175
The Future? 176

Appendix: The Animation Tutor™ DVD 179
Stephen K. Reed and Bob Hoffman

Dimensional Thinking 179
Brian Greer, Bob Hoffman, and Stephen Reed

Chemical Kinetics 180
Kathy Tyner, Stephen Reed, and Susan Phares

Personal Finance 180
Bob Hoffman and Stephen Reed

Population Growth 180
Stephen Reed, Bob Hoffman, and Diane Short

Average Speed 181
Stephen Reed, Jeff Sale, and Susan Phares

Catch Up 181
Stephen Reed and Bob Hoffman

Task Completion 181
Stephen Reed, Susan Phares, and Jeff Sale

Leaky Tanks 182
Stephen Reed, Susan Phares, and Jeff Sale

Encapsulated Summaries 183

Chapter 1: Images Versus Words 183
Chapter 2: Images Before Words 183
Chapter 3: Estimation 184
Chapter 4: Spatial Metaphors 185
Chapter 5: Producing Images 185
Chapter 6: Manipulating Images 186
Chapter 7: Viewing Pictures 187
Chapter 8: Producing Diagrams 187
Chapter 9: Comprehending Graphs 188
Chapter 10: Words and Pictures 189
Chapter 11: Vision and Action 189
Chapter 12: Virtual Reality 190

Chapter 13: Science Instructional Software 191
Chapter 14: Mathematics Instructional Software 192
Chapter 15: Conclusions 192

References 195
Author Index 209
Subject Index 215

Preface

Language is a marvelous tool for communication, but it is greatly overrated as a tool for thought. This is the opening sentence of Chapter 1 but it is worth repeating because it is why I wrote this book. *Thinking Visually* documents the many ways in which pictures, visual images, and spatial metaphors influence our thinking. It discusses both classic and recent research that supports the view that visual thinking occurs not only where we expect to find it, but also where we do not. Much of comprehending language, for instance, depends on visual simulations of words or on spatial metaphors that provide a foundation for conceptual understanding. Sometimes we are conscious of visual thinking. At other times, it works unconsciously behind the scenes.

I wrote this book to share cognitive science with an audience that is larger than that of cognitive scientists. The popularity of Malcom Gladwell's book *Blink* made me realize that many people are interested in cognition if its story could be told in an interesting manner. *Thinking Visually* tries to do this by

- including many pictures and diagrams for readers who like illustrations
- discussing visual thinking in children for parents interested in cognitive development
- presenting problems that require visual thinking for problem solvers
- encouraging educational applications for educators
- including the Animation Tutor™ DVD for students
- commenting on future changes in instruction

I am grateful for the inspiration provided by Malcom Gladwell in his books and in his address at the 2006 Convention of the Association for Psychological Science. This convention also provided me with another strong dose of motivation through The Mind in the Media symposium organized by Michael Gazzaniga. A distinguished group of panelists—Robert Bazell of *NBC News*, Daniel Henninger of *The Wall Street Journal*, Erica Goode of *The New York Times*, author Tom Wolfe, and William Safire of *The New York Times*—argued that if experimental psychologists desire increased recognition and funding of their research, they must convince the public that they are doing important work. I hope this book accomplishes that objective.

Now for my fellow cognitive scientists. I have not forgotten you. The theoretical inspiration for this book is the recent research on how embodied cognition influences thinking. Let me say up front, however, that *Thinking Visually* is not a monograph that explores this topic from all perspectives. It states a personal view, but one that has been shaped by many years of reading, researching, and reflecting about those activities that require thinking visually. I gained a tremendous amount of new knowledge from writing this book. I hope you will learn much new information from reading it.

Acknowledgments

People who influenced my career are tremendously important but so are those who prepared me for it. Three who particularly stand out are Mike Posner, Brendan Maher, and Mort Friedman. During my sophomore year at the University of Wisconsin six other undergraduates and I met weekly with Professor Posner to read and discuss articles in an honors seminar. Those articles, such as George Miller's on the limited capacity of short-term memory and the Petersons' on rapid forgetting, introduced me to the wonders of cognitive psychology. Brendan Maher's direction of my undergraduate thesis during my senior year taught me how to do research. Mort Friedman's support and guidance throughout my graduate study helped me turn my research into published articles.

I would also like to acknowledge those people who helped me with writing this book, beginning with Art Markman and Laura Novick who read the entire draft and provided many valuable suggestions. Others who read parts of the book include Kathie Hoxsie, Anita Marovac, Diana Osborne, and Joan Scheu. I am also grateful to instructor Lynne Friedmann and my fellow students in our Science Writing II course. Lori Handelman, my initial editor at Lawrence Erlbaum Associates, guided me through the initial stages and a team of experts at Taylor & Francis assured the high quality of the final product.

An important part of this book is the Animation Tutor™ DVD that we produced during a 4-year curriculum development project supported by the National Science Foundation. It illustrates an application of research on visual thinking to improve mathematical reasoning. Producing effective software requires a team of talented people and I have been fortunate to find them. Jeff Sale and Susan Phares did much of the initial programming in Macromedia Director. Bob Hoffman joined the team as software designer after completing the Mission Museum project described in Chapter 12. Brian Greer was the principle designer of the Dimensional Thinking module and Kathy Tyner was the principle designer of the Chemical Kinetics module. Diane Short contributed to the Population Growth module and graciously allowed us to evaluate the modules in her mathematics classes at Southwestern (community) College. Kien Lim, Nghiep Quan, Perla Myers, and Ian Whitacre provided valuable feedback on content.

My initial book published in 1973 was the first volume in the Academic Press Series in Cognition and Perception. Shortly after its publication, I met a young man who informed me that he had worked on the early stages of my book but then left Academic Press to build his own company. It is a privilege 37 years later to contribute one of the final volumes signed by that company. Countless others and I are very grateful to Larry Erlbaum and Lawrence Erlbaum Associates for the tremendous impact they have had on distributing knowledge.

1 Images Versus Words

Language is a marvelous tool for communication, but it is greatly overrated as a tool for thought. Because we are constantly exposed to language, we believe that thinking verbally dominates our lives. Thinking visually, if it occurs at all, is hiding in the shadows.

It is easy for me, as a psychology professor, to see why people would place much more emphasis on verbal thinking than on visual thinking. Universities have Departments of Linguistics. They do not have Departments of Visualization. Psychology curricula include courses in psycholinguistics. They typically do not include courses in visual thinking. There are many books on language. There are relatively few books on visual thinking.

Therefore, it should not be surprising that many books on thinking have emphasized verbal thought. Robert Sternberg at Tufts University is one of psychology's foremost experts on intelligence. His book with Talia Ben-Zeev, *Complex Cognition: The Psychology of Human Thought* (Sternberg & Ben-Zeev, 2001) provides an impressive coverage of many topics including concepts, knowledge representation, reasoning, problem solving, decision making, language and thought, human and artificial intelligence, creativity, expertise, development, and teaching. *Complex Cognition* is primarily a book about verbal cognition. The term *spatial visualization* in the subject index refers to only a single page, the term *visual* is used only as an adjective for the word *mask*, and the term *image* is not even listed in the index.

Yet, it is easy to generate examples of visual thinking if we pause to reflect. A baby who remembers that his mother hid his toy under a blanket is thinking visually. An investor who tracks the price of a stock on a graph to determine when to buy or sell is thinking visually. A scientist who studies the structure of molecules is thinking visually. An interior decorator who coordinates the colors in a room is thinking visually. A fashion designer who creates a new dress is thinking visually. An athlete who mentally simulates an action before executing it is thinking visually.

Some of the most impressive achievements of humanity have resulted from visual thinking. Einstein achieved his remarkable insights into the nature of space and time through mental simulations. He imagined himself traveling through space alongside a beam of light while viewing idealized physical bodies including clocks and measuring rods. Although Einstein's thought experiments are the most famous examples of the power of visualization in scientific discovery, there are many other examples of how scientific thinking involves visual thinking. High-powered visual thinking also occurred for other scientists such as James Watson and Francis Crick as they constructed molecular models of the structure of DNA. Words such as *insight* and *imagination* reflect the power of visual thinking.

Great architecture is also a product of thinking visually. One example, Fallingwater, is a remarkable home that was built in the woods of Pennsylvania.

The architect, Frank Lloyd Wright, cantilevered the decks of the home over a waterfall to make it look more like a modern home than one built in the 1930s (Figure 1.1). Wright's buildings were often more impressive for their visual form than for their function. The visual forms of Wright's designs are spectacular as revealed by the horizontal lines of the Robie House in Chicago, the sculptural look of the support columns of the Johnson Wax building in Racine, the Maya Temple design of the Hollyhock House in Los Angeles, and the spiral ramps of the Guggenheim Museum in New York.

The creative designs of the scientist and the architect came together for me in a memorable moment that occurred about 5 years ago when I attended a lecture in a newly constructed building for the Salk Institute in La Jolla, California. A famous architect named Louis Kahn designed the original buildings. Adding new buildings to the site was highly controversial because many people thought it would infringe upon Kahn's creation. As I was leaving the new building after the lecture, I had the opportunity to discuss this controversy with Francis Crick, the acting director of the Salk. It was a rare opportunity to hear one brilliant visual thinker talk about the work of another brilliant visual thinker. Crick explained that most of the initial critics were pleased with the outcome. Placing the new buildings some distance from the original ones had preserved the visual integrity of Kahn's creation.

You may believe by now that I have stacked the deck in favor of visual thinking by discussing the works of famous scientists and architects. So let us look at the contributions of visual thinking to a profession that best captures the creativity of *verbal* thinking—writing. One of the important talents of gifted writers is their ability to paint pictures with words, as illustrated by a description of a simultaneous sunset and moonrise across the Carolina marshes:

> Behind us the sun was setting in a simultaneous congruent withdrawal and the river turned to flame in a quiet dual of gold. The new gold of moon astonishing and ascendant, the depleted gold of sunset extinguishing itself in the long westward slide, it was the old dance of days in the Carolina marshes, the breathtaking death of days before the eyes of children, until the sun vanished, its final signature a ribbon of bullion strung across the tops of water oaks. The moon then rose quickly, rose like a bird from the water, from the trees, from the island, and climbed straight up—gold, then yellow, then pale yellow, pale silver, silver-bright, then something miraculous, immaculate, and beyond silver, a color native only to southern nights. (Conroy, 1986, p. 5)

Pat Conroy's sentences from *The Prince of Tides* come attached with visual images.

Other great writers use visual metaphor to make the abstract more concrete, as in the following description of a small town in Kansas:

> Until one morning in mid-November of 1959, few Americans—in fact, few Kansans—had ever heard of Holcomb. Like the waters of the river, like the motorists on the highway, and like the yellow trains streaking down the Santa Fe tracks,

Images Versus Words

FIGURE 1.1 Fallingwater designed by Frank Lloyd Wright.

drama, in the shape of exceptional happenings, had never stopped there. (Capote, 1965, p. 5)

This description of dramatic events traveling through space, looking for a place to land, indicates that Holcomb, Kansas is not a likely candidate. But in November 1959 a dramatic event did find Holcomb—and provided the story for Truman Capote's book, *In Cold Blood*.

Kaye Gibbons (2006) selects a different spatial metaphor for her book, *The Life All Around Me by Ellen Foster*. Gibbons begins a description of Ellen's mother with the statement "The hole was emptier than holes with merely nothing in them..." (p. 66). The power of this metaphor is that it has both an emotional and a cognitive impact. Its emotional impact is the suggestion of abandonment and hopelessness. Its cognitive impact is the dilemma of how one hole can be emptier than others. The complete sentence provides the answer: "The hole was emptier than holes with merely nothing in them, because it was missing everything that had been possible before." The hole seems emptier because once it was filled.

THINKING WITH SYMBOLS

If visual thought is so important, why do we place so much emphasis on verbal thought? Steven Pinker (1994) provides an answer in his book, *The Language Instinct*:

People can be forgiven for overrating language. Words make noise or sit on a page, for all to hear and see. Thoughts are trapped inside the head of the thinker. To know what someone else is thinking, or to talk to each other about the nature of thinking, we have to use—what else, words. It is no wonder that many commentators have trouble even conceiving of thought without words—or is it that they just don't have the language to talk about it? (p. 67)

Pinker concludes that people do not think in English, Chinese, Apache, or any other language. They think in a language of thought. He proposes that a language of thought probably looks a bit like other languages, consisting of arrangements of symbols that represent concepts. Before considering what these symbols might be, let us look first at what are symbols.

According to the University of Virginia psychologist Judy DeLoache (2004), a symbol is something that someone intends to represent something other than itself. Putting the word *something* not only once, but twice, into a statement makes the definition difficult to grasp but she insists that every component of her definition is essential. The word *something* implies that both the symbol and its referent (what it represents) can be anything—spoken words, printed words, pictures, video images, numbers, graphs, and an infinite list of other possibilities.

The word *someone* in DeLoache's definition typically points to humans because according to DeLoache (2004):

Although remarkable success has been achieved teaching non-human primates and some other animals to use certain symbols, the creative and flexible use of a vast array of different types of symbols is unique to humans. The emergence in evolution of the symbolic capacity irrevocably transformed our species, vastly expanding our intellectual horizons and making possible the cultural transmission of knowledge to succeeding generations. (p. 66)

Another important word in the definition is *represent*. Symbols refer to or denote something. They are not merely associated with their referents. This makes it difficult to determine whether a child's use or understanding of words is truly symbolic. For example, a young child who says "dog" when seeing a dog or a picture of a dog may only have formed an association based on repeated experience without realizing that the word can represent the object. It is not always easy, even for adults, to know what a symbol represents. What is the intended meaning of a body in the position of Leonardo Da Vinci's Vitruvian Man? Deciphering this mystery is the first of many puzzles faced by Harvard symbologist Robert Langdon in Dan Brown's (2003) novel *The Da Vinci Code*.

Finally, the word *intends* is a crucial part of the definition. Nothing is inherently symbolic. A symbol is created when someone uses it with the goal of denoting or referring to something else. DeLoache (2004) argues that a young child learns that a novel word is a label for an object only when another person is looking or pointing at the object. Pictures also become symbols through experience. A 9-month-old infant may place his lips on the nipple of a depicted baby bottle. By 18 months of age, children treat pictures symbolically as objects of contemplation and communication, rather than as objects to manipulate.

PROPOSITIONS

One of the problems with using language to represent thought is that language can be too flexible. Pinker (1994) illustrates the richness of language with the following four sentences:

Sam sprayed paint onto the wall.
Sam sprayed the wall with paint.
Paint was sprayed onto the wall by Sam.
The wall was sprayed with paint by Sam.

The four sentences describe the same event in different ways. You know that all four sentences have the same meaning, but to know this, there must be some underlying thought that is more basic than each of the sentences is. The central idea in all four sentences is that Sam sprayed paint and this caused paint to go onto the wall. This underlying thought can be expressed by propositions that are derived from logic and computer languages. Pinker (1994) shows how this central idea can be expressed by combining propositions that look something like:

(Sam spray $paint_i$) cause ($paint_i$ go to (on wall))

The subscript indicates that the paint going on the wall is the same paint that Sam is spraying. Propositions are popular among cognitive scientists because they represent meaning in a uniform notation, without the specific details used to express the meaning. Stripping away these details can have advantages, but it can also have disadvantages. For example, the initial word in the sentence may convey its importance. The emphasis is on *who* is spraying in the first two sentences, *what* is being sprayed in the third sentence, and *which* object is being sprayed in the fourth sentence. This emphasis is lost when all four sentences are represented by the same set of propositions.

Propositional representations also omit the source of the information. You read about Sam painting the wall, but that information is not contained in the propositional representation. It does not specify whether you obtained the information by reading, hearing, or seeing Sam paint the wall.

Another disadvantage of propositional representations is that, unlike words and images, they are models of reality, not reality. They enable cognitive scientists to model people's knowledge in the same way that mathematical equations enable physicists to model interactions among physical objects. But if people do not use propositional structures in their thinking, what kinds of symbols do they use? One answer, of course, is words. I am not arguing that words are seldom used; only that visual symbols are equally important. This importance is recognized in a theory that emphasizes perceptual symbols.

PERCEPTUAL SYMBOLS

We can all agree that language forms a powerful symbol system that has greatly expanded our intellectual capabilities. However, this emphasis on language may have distracted us from recognizing that our perceptual experiences also form the basis for a powerful symbol system. Most of the knowledge structures proposed by cognitive scientists have stripped away our initial perceptual experiences to represent knowledge in a rather abstract form, such as the propositions discussed in the previous section. As pointed out by Wilson (2002),

> Traditionally, the various branches of cognitive science have viewed the mind as an abstract information processor, whose connections to the outside world were of little theoretical importance. Perceptual and motor systems, though reasonable objects of inquiry in their own right, were not considered relevant to understanding "central" cognitive processes. Instead, they were thought to serve merely as peripheral input and output devices. (p. 65)

Perceptual experiences come in different sensory modalities based on audition, vision, taste, smell, touch, and movement. However, these differences are ignored by amodal theories that propose we store our experiences in memory independent of these modalities by using the same abstract description for all experiences. A dramatically different approach is the perceptual symbols theory proposed by Barsalou (1999). It is a perceptual theory of knowledge in which perceptual experiences are directly stored in memory and can therefore form the basis for visual thinking.

Figure 1.2 shows this distinction, as is illustrated for the concept *car*. Amodal symbol systems originate in perceptual experiences, but are represented in memory by nonperceptual knowledge structures such as a feature list, semantic network, or frame. A feature list describes the features of a car such as having an engine and wheels. A semantic network shows its relation to other concepts such as *vehicle* and *boat*. A frame specifies the feature values of a particular car such as having a 4-cylinder engine and being red. In contrast, perceptual symbols are retrieved by reenacting or simulating perceptual experiences. A person could imagine seeing a red car or listening to the sound of its engine.

The simulation hypothesis is simply that the same neurons that produce the initial perceptual experience can be reactivated to produce a reenactment of that experience. Many neuroscientists have provided compelling support for perceptual simulation through neuroimaging studies. The growth of cognitive neuroscience has contributed greatly to our understanding of cognitive processes by identifying which parts of the brain are used to perform a variety of cognitive tasks. Brain imaging techniques such as positron-emission tomography (PET) and functional magnetic resonance imaging (fMRI) measure cerebral blood flow by sensing either low-level radiation (PET) or a magnetic signal (fMRI). Increased blood flow is required by those parts of the brain that are working the hardest.

Images Versus Words

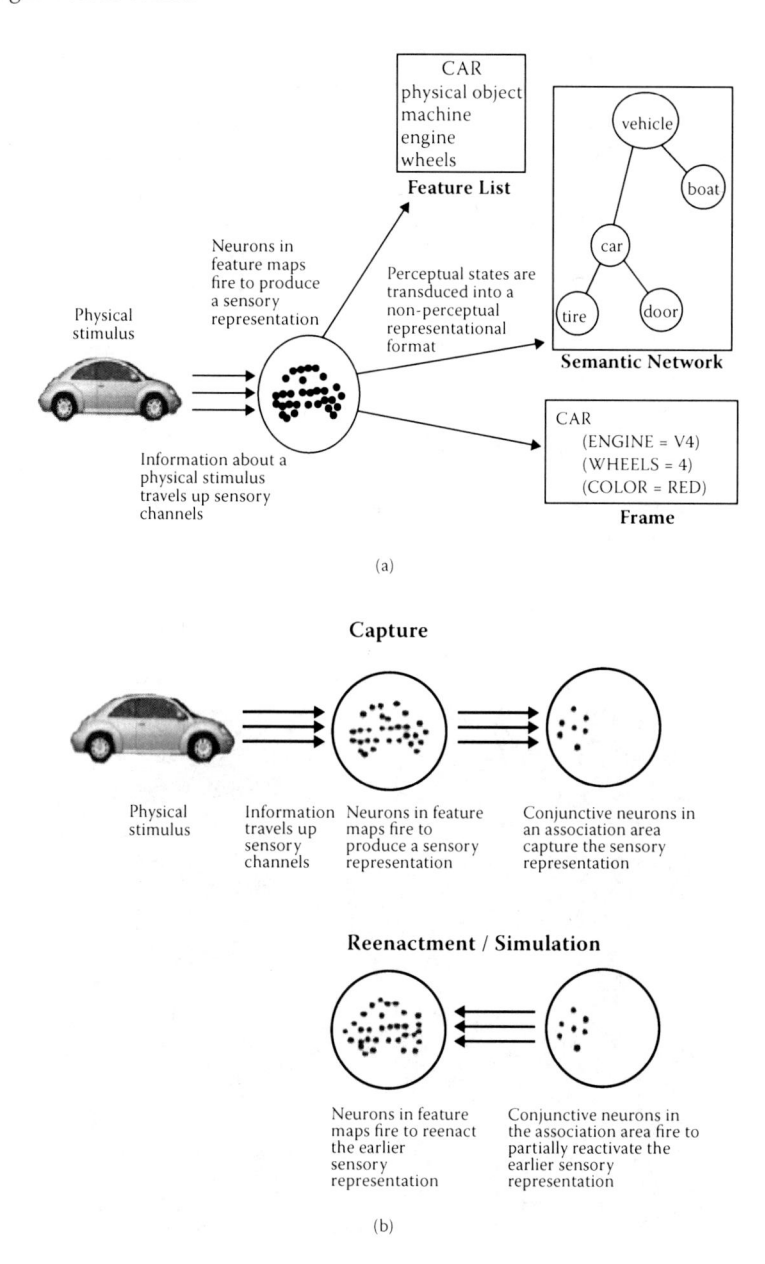

FIGURE 1.2 Representation of information in (a) amodal symbol systems and (b) perceptual symbol systems. (From "Grounding Conceptual Knowledge in Modality-Specific Systems," by L. Barsalou, W. K. Simmons, A. K. Barbey, & C. D. Wilson, 2003, *TRENDS in Cognitive Sciences, 7*, 84–91. Copyright 2003 by Elsevier. Reprinted with permission.)

Brain imaging studies have demonstrated that both perceiving a visual stimulus and imaging a visual stimulus increase blood flow in the visual cortex.

Figure 1.3 shows the activation of the visual cortex in the back of the brain as measured by fMRI. Figure 1.3a illustrates strong activation when a person viewed a flashing light pattern. Figure 1.3b shows weak activation when a person imagines the flashing light pattern. Figure 1.3c shows moderate activation when a person imagines walking in his or her hometown. The activation of the visual cortex is not as strong for imagination as it is for perception—an important difference that I will discuss in Chapter 5.

It should not surprise us that imagining a light pattern or a walk in familiar surroundings provides evidence of visual imagery. However, there is increasing evidence that visual simulation can also help us understand language. Your visual imagination was perhaps stimulated when you read Conroy's evocative description of the play of light on the Carolina marshes. But your comprehension of even the mundane statement "Sam sprayed paint onto the wall" may have been aided by visually simulating the event. Results obtained by Stanfield and Zwaan (2001) support the hypothesis that visual simulations help us comprehend the meaning of such verbally described events.

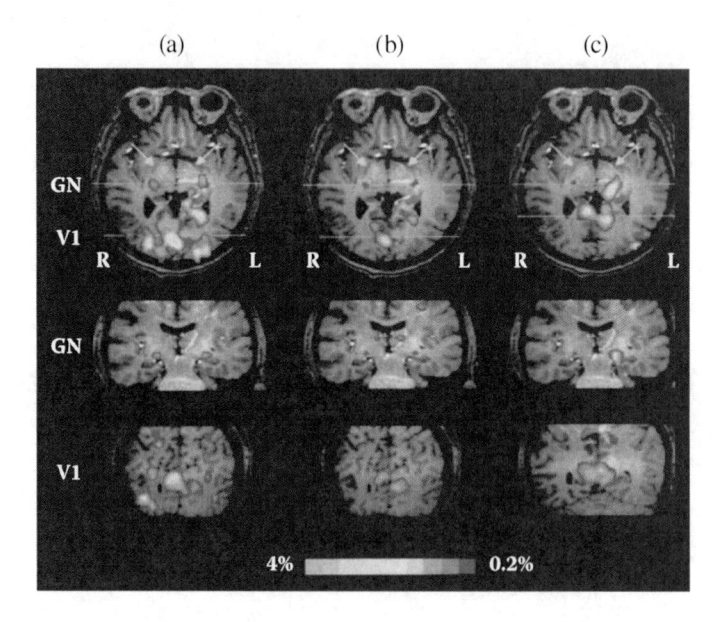

FIGURE 1.3 Activation of the visual cortex (V1) occurs when people (a) view a flashing light pattern, (b) imagine the flashing light pattern, or (c) imagine walking in their hometown. (From "Human Primary Visual Cortex and Lateral Geniculate Nucleus Activation during Visual Imagery," by W. Chen, T. Kato, X. H. Shu, S. Ogawa, D. W. Tank, & K. Ugurbil, 1998, *Neuroreport, 9*, 3669–3674. Copyright 1998 by Lippincott, Williams, & Wilkins. Reprinted with permission.)

Images Versus Words

Stanfield and Zwaan (2001) hypothesized that visual simulations of verbal statements should include an object's orientation. Mentioning that I put a pencil in the drawer should evoke an image of a horizontal pencil, and mentioning that I put a pencil in the cup should evoke an image of a vertical pencil. Stanfield and Zwaan tested their hypothesis by asking students at Florida State University to decide quickly whether a pictured object had been mentioned in a sentence that they just read. You can approximate an idea of this task by responding if the object below is mentioned in each of the three sentences in the Object Verification Task. This demonstration is approximate because the test object did not appear until after participants read the sentence in Stanfield and Zwaan's experiment.

Object Verification Task

1. She pounded the nail into the floor.
2. She painted the fence with a brush.
3. She pounded the nail into the wall.

According to the visual simulation hypothesis, the time to verify a mentioned object should depend on whether the picture matches the implied orientation in the sentence. The results supported the hypothesis, as illustrated by the distinction between the two sentences "She pounded the nail into the floor" versus "She pounded the nail into the wall." The readers were faster in confirming a picture of a vertical nail following the first sentence (pounding a nail into the floor) and were faster in confirming a picture of a horizontal nail following the second sentence (pounding the nail into the wall).

Other research by Zwaan and Yaxley (2003) indicates that visual simulations influence decisions about whether two words are semantically associated. You can participate in this task by judging whether the vertically aligned word pairs in the Word Association Task are related in meaning. Try to make your judgment for each of the pairs as quickly as you can.

Word Association Task

Judge whether the two words in each vertically aligned pair are associated.

ATTIC	LAKE	SWAN	CANDLE	TAXI	MOUTH
BASEMENT	BOAT	SINK	FLAME	PEACH	NOSE

You might not have noticed any differences in the time it took you to make these judgments, but precise measures of people's response times indicate that there are differences. If people use visual images to help them make the decision, the position of the words should influence response times. The results supported this prediction. When the vertical alignment of the words matched the visual alignment in people's images (such as ATTIC above BASEMENT), they were significantly faster in verifying semantic relatedness than when the alignments mismatched (BASEMENT above ATTIC). You should have been faster in judging that MOUTH and NOSE are related in meaning than if I had placed the word NOSE above the word MOUTH.

Barsalou's (1999) article in *Brain & Behavioral Sciences* has provided the impetus for much of the current research on perceptual simulations of thought. However, other cognitive scientists have proposed similar views. In his presidential address to the American Association for Artificial Intelligence, Randall Davis (1998) spoke about the evolution of intelligence. He concluded his talk by considering what parts of intelligence might usefully be explored more thoroughly. Davis proposed that as an alternative to the view that thinking is a form of internal verbalization, thinking should be viewed as reliving our perceptual and motor experiences. He admitted that these ideas were speculative. They are no longer.

ORGANIZATION OF THE BOOK

This book is about visual and spatial thinking. I will provide much evidence that spatial thinking (reasoning about objects in one, two, or three dimensions) involves visual thinking but I will occasionally talk about spatial thinking without providing the evidence. I believe that most of these cases also involve visual thinking but the research has not always been done.

Let me make the distinction between visual and spatial thinking with a common example. Imagine that you ask for directions and are told that you need to travel five blocks on Elm Street, then turn right on State Street, travel three blocks on State Street, and then turn left at the church onto Pearl Street. This spatial task provides an opportunity for visual thinking, but does not require it. If you are a visual thinker, you may have already formed a visual image of a map that will help you find your destination (Brunye & Taylor, 2008). If you are not a visual thinker, you could simply memorize the verbal instructions.

Constructing a mental map can have memory benefits because, as we will see in Chapter 5, a visual image provides an additional means of remembering. It can also help on your return because following the verbal directions requires reversing the order and the left–right turns. You would need to turn *right* at the church onto State Street and then *left* onto Elm Street. Going to and from a destination can be accomplished without visual thinking but I have found that it is easier when I can construct and scan a mental map.

It can be a challenge for psychologists to determine the relative amount of visual and verbal thinking but this challenge is made easier by studying babies'

Images Versus Words   11

accomplishments before they acquire language. Chapter 2 ("Images Before Words") explores these early stages of perceptual and conceptual development before the acquisition of language. Chapter 3 ("Estimation") discusses the spatial and numerical abilities needed to perform quantitative reasoning, and the relation between qualitative and quantitative reasoning. Chapter 4 ("Spatial Metaphors") examines how spatial metaphors help us understand abstract ideas.

The next two chapters explore visual imagery. The emergence of research on static images is the topic of Chapter 5 ("Producing Images"). Chapter 6 ("Manipulating Images") focuses on dynamic images in which people create products by combining parts, "running" mechanical systems, and inducing rules. Both chapters examine the limitations of visual images, and Chapter 5 shows how these limitations help us distinguish between reality and imagination.

The next three chapters are on visual displays. Chapter 7 ("Viewing Pictures") asks why a picture is worth a thousand words. It includes discussion on both the instructional and emotional impact of pictures. Chapter 8 ("Producing Diagrams") examines how students select a visual representation of a problem and how students and experts generate their own diagrams to represent problems. Chapter 9 ("Comprehending Graphs") focuses on the crucial role of knowledge in interpreting graphs and the role of instruction in aiding this interpretation.

Although this book's topic is visual cognition, visual cognition by itself would be of limited value. It needs to be integrated with language, concepts, and action to have a major intellectual impact. Chapter 10 ("Words and Pictures") reviews research and theory on the integration of words and pictures. Chapter 11 ("Vision and Action") examines the role of action in learning. Action and visualization are intertwined when manipulating either physical or virtual objects. The integration of vision with action and other modalities is exploited in the design of virtual reality, discussed in Chapter 12 ("Virtual Reality").

The final three chapters on instructional animation show how animation-based software can support visual thinking in science and mathematics education. Chapter 13 ("Science Instructional Software") provides an overview of exemplary animation software that supports visual reasoning for a variety of curricula including biology, chemistry, complex systems, and physics. Chapter 14 ("Mathematics Instructional Software") includes a discussion of the SimCalc, Animation Tutor, and ANIMATE projects. All three use animation and virtual manipulation to encourage visual thinking. Chapter 15 ("Conclusions") argues that we need to do more to create and evaluate software that encourages and supports visual thinking.

SUMMARY

The important role of language in communication, at times, has caused us to neglect the role of vision in thinking. However, there are many examples of the importance of visual thinking in fields such as architecture, design, and science. There is also recent evidence that visual simulation is important in understanding language and in verbal reasoning. For example, people are faster in verifying that a pictured object was mentioned in a sentence if the object's orientation matches the orientation implied by the sentence. They are faster in verifying that two words are associated (such as flame and candle) if their vertical alignment matches their relative position in an image. These and many other findings have resulted in the view that much of thinking can be viewed as reliving our perceptual and motor experiences, rather than as a form of internal verbalization. This book provides an overview of research on the many ways we think visually and discusses the educational implications of this research.

2 Images Before Words

Our cognitive abilities to comprehend, remember, reason, solve problems, and make decisions depend on a rich combination of images and words. This makes it difficult for psychologists to separate the relative contributions of language and imagery. However, we have an ace up our sleeve—we can study babies who have images before they have words. In addition, according to the testimony of proud parents, their children exhibit many marvelous mental accomplishments before they acquire language. Developmental psychologists have studied some of these children and have found that most of these claimed achievements actually occur.

For example:

- A 6-month-old infant remembers whether she was shown, 2 weeks earlier, the face of a baby, a man, or a woman.
- A 9-month-old infant generalizes concepts by using small toys to show animals, but not vehicles, drinking.
- A 10-month-old infant chooses the bucket containing more crackers after two crackers were placed in one and three crackers in the other.

This chapter reviews, and then builds on, these findings to show how spatial ability provides a conceptual foundation that is needed for language.

VISUAL MEMORY

Visual thinking builds on our abilities to perceive and remember visual patterns. Robert Fantz (1961) is responsible for much of the early research on infants' visual perception. Fantz developed a looking chamber that was placed above the infant as she lay on her back. An observer placed two visual patterns on the ceiling of the chamber and watched through a peephole to record how much time the baby looked at each pattern.

One of Fantz's early experiments studied infants ranging in age from 4 days to 6 months. The patterns consisted of (a) a cartoonish face, (b) a scrambled cartoonish face with rearranged features, and (c) a plain pattern consisting of two solid patches of color. The babies viewed all possible pairs of patterns: a with b, a with c, and b with c, during a sequence of 2-minute tests. Babies of all ages showed a slight preference for looking at the cartoonish face than the scrambled face. The plain pattern finished a distant third.

Fantz and I became colleagues when I joined the Psychology Department at Case Western Reserve University in 1971. Another colleague, Joseph Fagan, had recently extended Fantz's procedure to study the ability of babies to remember visual patterns. Fagan's (1973) research depended on the previous discovery that infants prefer to look at new patterns. Show a baby two patterns, one of which he has already seen, and he will prefer to look at the pattern he has not seen. This, of course, presumes that the baby remembers the familiar pattern. If he forgets, both patterns will seem new and there should no longer be a preference for one pattern over the other. In one of these experiments, 21- to 25-week-old infants looked at a facial photograph of a man, a woman, or a baby. They were later shown the familiar face paired with a new face that they had not previously seen. Even after a 2-week delay, infants spent significantly more time viewing the new face, demonstrating that they still had memory of the familiar face.

Fagan's findings demonstrated that infants could remember enough about a previously seen face to later discriminate it from a new face. However, the discrimination task was relatively easy because it required discriminating among the faces of a man, a woman, and a baby. About the same time, Ed Cornell, a student of Fagan's was collecting data for his doctoral dissertation on infant's discrimination of the faces shown in Figure 2.1. Would babies who saw the faces of four different men or women on each trial spend more time looking at faces than babies who saw the same man or woman on each trial? One group of babies viewed either the four women or the four men shown in Set 1. Another group of babies viewed either the four perspectives of the woman's face or the four perspectives of the man's face shown in Set 2. A third group of babies always viewed the same woman or the same man shown in Set 3.

Cornell (1974) measured how much time the babies spent looking at the faces during six 10-second viewing trials that were separated by 4 seconds. He found that 23-week-old infants spent slightly more than 8 of the 10 seconds viewing the faces during the first trial. Those in the first condition maintained their attention to faces over the next five trials because the faces varied among the four women or among the four men. Babies in the second condition attended less to the different perspectives of the same face because the faces appeared more similar and therefore less interesting. Babies in the third condition attended the least as they repeatedly saw the same face. In contrast, 19-week-old infants did not show differences among the three conditions. All the men's faces and all the women's faces apparently looked the same to them so their interest was not maintained by showing variations of the faces.

FROM MEMORY TO CONCEPTS

This research by my colleagues at Case Western Reserve demonstrated that babies have preferences for some visual patterns over others (Fantz, 1961) and, by 6 months of age, can differentiate (Cornell, 1974) and remember (Fagan, 1973) the faces of strangers. The ability of babies to recognize objects such as faces provides a foundation for visual thinking because visual thinking often depends on the ability to remember and manipulate visual images. In addition, memory for objects forms the basis for forming concepts that are more general than specific objects. Concepts are also helpful for thinking visually.

Images Before Words

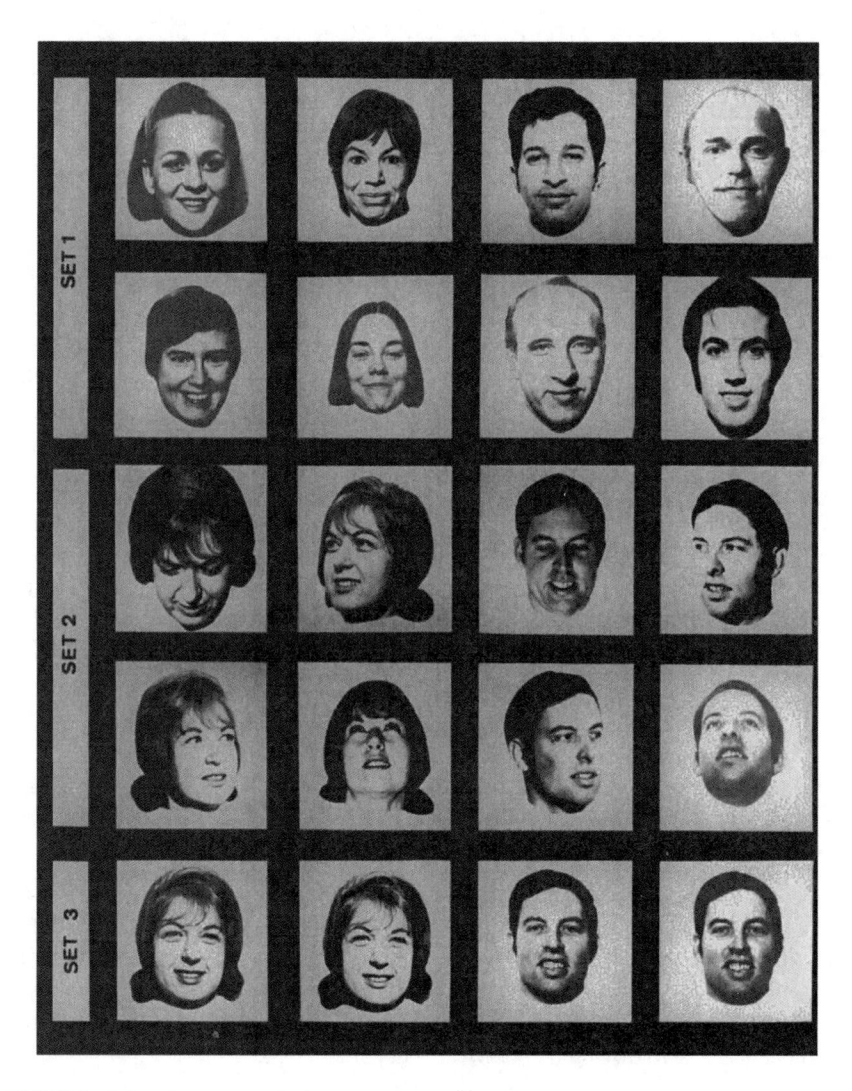

FIGURE 2.1 Example of faces used in Cornell's experiment. (From "Infants' Discrimination of Photographs of Faces Following Redundant Presentation," by E. Cornell, 1974, *Journal of Experimental Child Psychology, 18*, 98–106. Copyright 1974 by Elsevier. Reprinted with permission.)

So how do babies progress from objects to concepts? Jean Mandler (2004), a developmental psychologist at the University of California in San Diego, provides some promising answers in her book, *The Foundations of Mind: Origins of Conceptual Thought*. I will summarize those parts of Mandler's theory that are most relevant to understanding the visual foundations of thinking.

Mandler's theory about the origins of conceptual thought was stimulated by the writings of the famous Swiss psychologist, Jean Piaget. She represents his theory in the following way:

> One of the most prevalent intuitions about early cognitive development is that young infants have virtually no conceptual life. Instead, babies are described as sensorimotor creatures who understand the world solely through their perceptual and motor systems. They recognize things they have seen before, they can move themselves and manipulate objects, but they have no concepts and so cannot think, recall the past, or imagine the future. (Mandler, 2004, p. 3)

Mandler's own research eventually led her to a very different view in which, by a few months of age, infants function in ways that continuously evolve into the thinking that later characterizes older children and adults. In contrast to Piaget's (1977) view that children first enter a sensorimotor stage of development and only much later develop abstract thought, Mandler (2004) proposes that perceptual and conceptual development work in tandem. Babies form concepts, generalize from their experiences based on those concepts, and can recall absent objects and events. This conceptual system gives meaning to what they see and will later help them acquire language to talk about what they see.

A study by Mandler and McDonough (1996) illustrates one of the research techniques designed to measure babies' ability to form concepts. They modeled an event, such as giving a dog a drink from a cup, by using small replicas. They then gave 9- to 14-month-old infants the cup and substituted two other objects for the dog (such as a bird and a car) to see which objects the baby would use to imitate drinking. The infants used animals, but not vehicles, to imitate animal-specific behaviors such as drinking and sleeping. Using this and other research paradigms, Mandler and McDonough were able to show that infants initially acquire global concepts. They have an idea what an animal is, but are not sure how to differentiate among animals. They have an idea what a container is, but are not sure how to distinguish between a cup and a pan.

Mandler's theory of conceptual development has been greatly influenced by cognitive linguists who are interested in the underlying basis of the concepts expressed in language. Their basic claim is that one of the foundations of concepts (with or without language) is the formation of image-schemas that represent objects and events spatially at an abstract level. A helpful image-schema for thinking about a cup is containment, which describes a fully or partially enclosed space, without specifying the details of the container. The *containment* image-schema enables an infant to place a cup and a pan into the same functional category, but does not yet provide enough information to distinguish between them. Another example, the *path* image-schema, describes an object following a trajectory through space, without regard to the details of the trajectory or the characteristics of the object.

Images Before Words

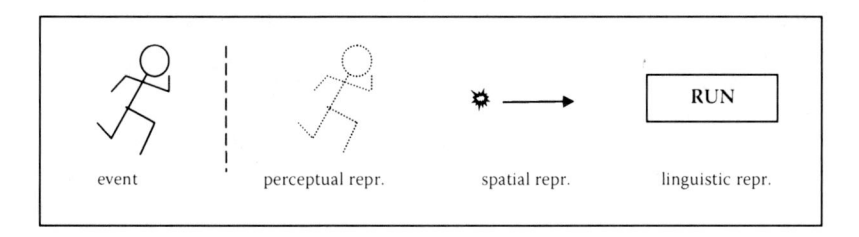

FIGURE 2.2 From perception to language. (From "Language and Space: Some Interactions," by A. Chatterjee, 2001, *TRENDS in Cognitive Sciences, 5*, 55–61. Copyright 2001 by Elsevier. Reprinted with permission.)

There are many similarities between the image-schemas proposed by Mandler (2004) and the perceptual symbol systems proposed by Barsalou (1999). Both perceptual symbols and image-schemas are extracted from perception and stored permanently in long-term memory. The main difference, according to Mandler, is that perceptual symbols are not as clearly differentiated from perception as are image-schemas. According to the perceptual symbols theory, we understand the sentence "The bird drank from the cup" through perceptual simulation. We can visually imagine this event. But image-schemas provide meaning to the perceptual symbols by specifying that a bird is animate so it makes sense for it to drink. They also provide category information such as some objects can serve the function of a container.

Figure 2.2 shows the transition from perception to language that Chatterjee (2001) uses to illustrate the evolvement of mental representations. The left figure represents the event "run" by the *perception* of a runner. The next figure represents a *mental simulation* of the runner. The broken lines indicate that our images are seldom as vivid or detailed as our perception. The *spatial schema* abstracts a simplified form that moves. Here is where we need Mandler's (2004) image schemas to distinguish the animate act of running from the inanimate act of moving. The *conceptual structure* in Chatterjee's figure (running is an event) is more like language in conveying information. Finally, the *verbal representation* specifies the word "run" to represent the act of running. This transition makes the important point that images come before words and provide mental objects and concepts that later can be linked to words. This transition from perception through spatial representation to language is also central in representing quantities.

SMALL QUANTITIES

As babies grow older, they can use visual memory to keep count of a small number of hidden objects. In a study by Feigenson, Carey, and Hauser (2002) 10- and 12-month-old infants watched an experimenter hide two crackers in one bucket and three crackers in another bucket. Eighty percent of the babies then chose the bucket with three crackers. However, they chose randomly when the experimenter hid four crackers in one bucket and one cracker in the other. The babies' memory limit was three objects.

Hauser's (2000) research at Harvard shows that rhesus monkeys have the same upper limit. A favorite food item, such as an eggplant, is placed behind a screen. Then another eggplant is added. The monkey looks only briefly after the screen is lifted because there are two eggplants, as expected. The experimenter then repeats the procedure but removes the second eggplant without the monkey's knowledge. The longer looking time at the unexpected single eggplant suggests that the monkey is trying to figure out what happened. The same puzzlement and longer looking time occurs when the monkey sees three eggplants, when expecting only two. However, the monkeys lose track of how many items are behind the screen when there are more than three.

The ability of monkeys and infants to keep count of a small number of objects forms a foundation for more sophisticated numerical reasoning in human adults. As argued by Hauser (2000):

> Today, while sitting in mathematics classes or perusing library bookshelves, we can study trigonometry, algebra, calculus and set theory. These systems showcase the endless creativity of the human mind and its invention of symbolic notation. We must not forget, however, that such systems stand on a foundation left behind by our animal ancestors. (p. 151)

SPACE, QUANTITY, AND TIME

Hauser's conclusion is consistent with the premise of a very readable book that appeared several years earlier. *The Number Sense: How the Mind Creates Mathematics* was written by Stanislas Dehaene, a French mathematician turned cognitive neuroscientist. The best single-sentence summary of the book is contained inside its jacket:

> Using clever experiments with animals, young infants, brain-lesioned patients, and high-tech imaging tools, psychologists have reached an amazing conclusion: Evolution has endowed each of us with an innate ability for arithmetic, and intuition of numerical quantities which, combined with our uniquely human ability for language, stands at the core of our ability to create mathematics. (Dehaene, 1997)

Let me discuss Dehaene's findings within the context of a theory of magnitude proposed by Vincent Walsh at the Institute of Cognitive Neuroscience, University College, London. Walsh's (2003) theory makes the following claims:

- Space, quantity, and time are linked by a common metric for action.
- The apparent specializations for time, space, and quantity develop from a single magnitude system operating from birth.
- The inferior parietal cortex is the location of the common magnitude system.
- Hemispheric differences in processing time, space, and number have

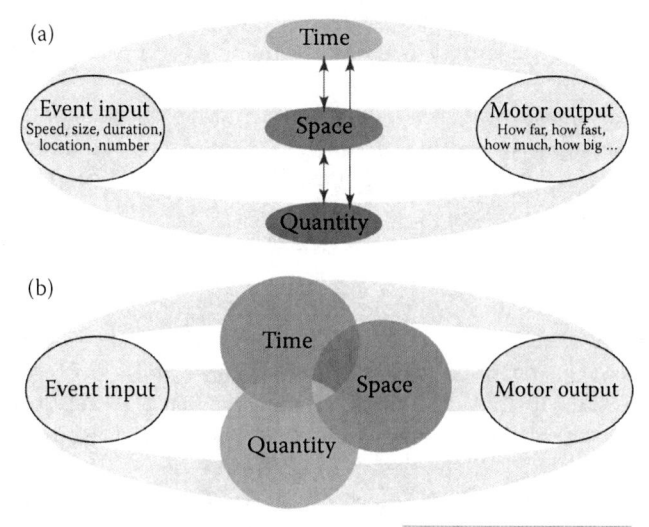

Images Before Words

FIGURE 2.3 Schematic representation that contrasts (a) a separate with (b) a common magnitude system that was proposed by Walsh. (From "A Theory of Magnitude: Common Cortical Metrics of Time, Space, and Quantity," by V. Walsh, 2003, *TRENDS in Cognitive Sciences, 7,* 483–488. Copyright 2003 by Elsevier. Reprinted with permission.)

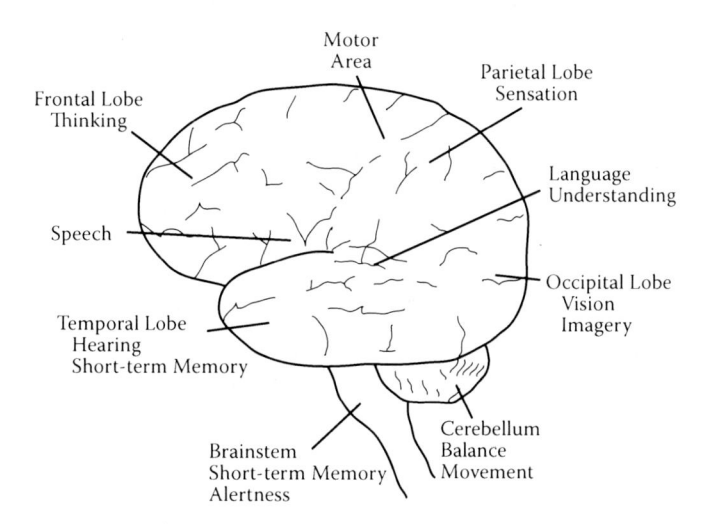

FIGURE 2.4 Major subdivisions of the left hemisphere of the cerebral cortex, with a few of their primary functions. (From *Applied Cognitive Psychology: A Textbook,* by D. J. Herrman, C. Y. Yoder, M. Gruneberg, and D. G. Payne, 2006, Mahwah, NJ: Lawrence Erlbaum Associates. Copyright 2006 Lawrence Erlbaum Associates. Reprinted with permission.)

emerged as a consequence of the fact that exact calculations require access to language in a way the spatiotemporal processing does not.

The interrelations among space, quantity, and time are represented in Figure 2.3 by the overlapping circles. Figure 2.4 shows the four lobes of the cortex, including the parietal lobe that Walsh (2003) identifies as the location of the common magnitude system.

There is both behavioral and neurological evidence to support Walsh's claims. You can try to convince yourself that *quantity and space* are closely related by performing the task at the bottom of Figure 2.5. Without counting, judge whether there are more white dots or black dots. Dehaene (1997) points out that these kinds of judgments depend upon the spatial display of the objects such as their density, their regularity, and the amount of area they occupy. Most people incorrectly believe that there are more white dots, perhaps because they are more tightly grouped.

Now try another of Dehaene's tasks by responding quickly to the numbers in the Number Comparison Task. In the first row, tap your left finger for each

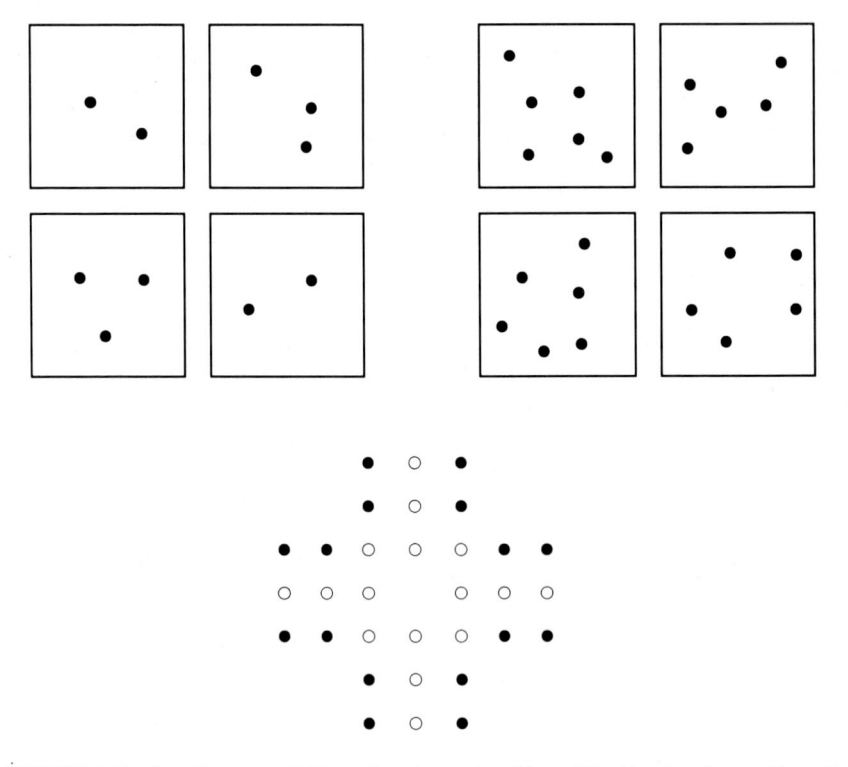

FIGURE 2.5 Dot displays used in estimation tasks. (From *The Number Sense: How the Mind Creates Mathematics*, by S. Dehaene, 1997, New York: Oxford University Press. Copyright 1997 by Oxford University Press. Reprinted with permission.)

number that is less than 5 and your right finger for each number that is greater than 5. In the second row, tap your left finger for each number that is less than 65 and your right finger for each number that is greater than 65.

Number Comparison Task

1 9 2 3 9 3 4 7 6 8 2 1 8 7 4 6

73 67 47 15 99 49 27 54 64 88 25

Do you think your responses were equally fast or were you influenced by how close the numbers were to 5 (first row) or to 65 (second row)? Research has demonstrated that people are faster when there is a large difference between the two numbers. It is easier to decide that 99 is larger than 65 than it is to decide that 67 is larger than 65. This task was initially studied by Moyer and Landauer (1967), who argued that people think about precise digits in terms of approximate magnitudes. The greater the *differences* in these magnitudes, the faster people are in judging which number is larger.

This finding is consistent with the hypothesis that we think of numbers spatially as occupying positions along a mental number line. The closer two numbers are on the line, the harder it is to tell which number is larger, just as it would be harder to tell which of two sticks is longer if there were a small difference in their lengths. In contrast, a verbal counting model might predict that we should be very fast in determining that 66 is larger than 65 because it immediately follows 65 when we count.

Dehaene (1997) has found other evidence to support the concept of a mental number line. Do the number comparison task again but this time tap your right finger for numbers less than 5 (or 65) and your left finger for numbers that are greater than 5 (or 65). You will likely found this task much more difficult. The reason is that the mental number line is ordered from low numbers on the left to higher numbers on the right so it is more natural to tap the left finger for lower numbers and the right finger for higher numbers.

The close link of *space and time* should be evident from our daily experiences. For instance, it usually takes more time to travel a longer distance than a shorter distance. We can often use one quantity to estimate the other. I use time to estimate space when I make coffee. I fill an opaque container with water so I cannot judge volume but I know how long to run the faucet to obtain 3 cups. I use space to estimate time when I cross a busy intersection on campus. It is a 3-minute wait between walk signs so I walk more rapidly to the intersection when there is a long line of cars waiting for the green light. It usually means that the green light and walk sign will soon appear.

The use of the same words for distance and time is another source of evidence. The statement "We don't know what lies ahead" can refer to a journey either through space or time. "A short trip" could refer to either time or distance. The sharing of words for space and time occurs for extent (long/short), ordering (ahead/behind), and proximity (near/far). Research by Casasanto & Boroditsky (2008) has revealed that people are unable to ignore irrelevant spatial information such as length when making judgments about duration. Space rather than time, therefore, appears to provide the foundation for constructing the links between the two concepts.

NUMERICAL REASONING WITH LANGUAGE

Numerical reasoning without language can take us only so far. Many aspects of mathematics require language, as is evident from the answers given by children in response to the question "Why do we study math in school?" Box 2.1 shows typical responses, which were collected by Catherine Valentino, a consulting author to Silver Burdett Mathematics. Language enables Maryanne to keep track of page numbers in a book, Michael to count, Raji to determine whether he is rich when he grows up, and Corky to realize that his teacher is not rich.

Language also helps us perform numerical calculations. Research by Dehaene, Spelke, Pinel, Stanescu, and Tsivkin (1999) has established that doing exact arithmetic requires language, but doing approximate arithmetic does not. They trained eight bilingual Russian–English speakers on twelve sums of two-digit numbers, totaling between 47 and 153. The sums were presented on a computer screen in either English (fifty-one plus thirty-five equals

---

BOX 2.1 ANSWERS GIVEN BY CHILDREN TO THE QUESTION "WHY DO WE STUDY MATH IN SCHOOL?"

Because it's hard and we have to learn hard things at school. We learn easy stuff at home like manners. Corrine – kindergarten.

Because all the calculators might run out of batteries or something. Thomas – Grade 1.

Because you have to count if you want to be an astronaut. Like 10..9..8…. blast off! Michael – Grade 1.

Because you could never find the right page. Maryanne – Grade 1.

Because when you grow up you couldn't tell if you are rich or not. Raji – Grade 2.

Because my teacher could get sued if we don't. That's what she said. Any subject we don't know – Wham! She gets sued. And she's already poor. Corky – Grade 3.

eighty-six) or in Russian to determine if the participants would be equally fast in selecting the correct sum when they were tested in their other language. The bilingual speakers were faster in selecting the correct answer when the test problems were given in the same language as occurred during training. However, switching languages did not make a difference when the bilinguals selected which of two alternatives was the *closest* to the correct answer in an approximation task.

The use of functional magnetic resonance imaging (fMRI) confirms that only exact calculations are dependent on language (Dehaene et al., 1999). This experiment measured brain activation during both exact addition and approximate addition. As in the previous experiment (Dehaene et al., 1999), exact addition required selection of the correct answer and approximate addition required selection of the closest answer. For instance, exact addition required determining whether 9 or 7 was the answer to 4 + 5 and approximate addition required determining whether 8 or 3 was an approximate answer to 4 + 5. The most active area of the brain for exact addition was the left inferior prefrontal cortex that is involved in word associations. In contrast, the most active area of the brain for approximate addition was the bilateral area of the parietal lobes that is involved in visuospatial reasoning. This is the area of the brain that Walsh (2003) targets as the location of a common magnitude system.

SUMMARY

The fact that images come before words provides cognitive scientists with the opportunity to study the perceptual, memory, and conceptual capabilities of babies before they acquire language. The visual system lays the foundation for objects and concepts that can later be described in words. It creates discriminable patterns such as faces, categories such as animals and vehicles, and image schemas such as containment and path. The visual system also provides an approximate sense of large quantities and an exact count of small quantities. The acquisition of language, however, enables us to go beyond a sense of approximation to describe relations among numbers and to perform exact calculations. We will find more evidence in the next chapter for the hypothesis that such calculations build on the spatial representations that support approximation.

3 Estimation

An approximate sense of quantity is very helpful for making estimates such as the time it will take to paint a fence. Many estimates involve measurement in which we can check our accuracy by using a tool such as a clock, a ruler, a scale, or a thermometer. An important attribute of these physical devices is that the numbers are equally spaced. The difference between 75 and 80 minutes is the same as the difference between 15 and 20 minutes. The difference between 23 and 25 meters is the same as the difference between 37 and 39 meters.

Imagine trying to measure distances using a ruler like the one shown in Figure 3.1. It is difficult to imagine but that is exactly the ruler that many young children mentally use before they learn to put equal spaces between successive integers. Why would young children use such a strange ruler? The reason seems to be that children's initial number line is based on their perceptual experiences. You can experince one of these effects for yourself by rapidly judging which square has more dots for seach of the four horizontal pairs on Figure 2.5. It is easier to distinguish between two and three dots than between five and six dots. It is more challenging to detect the one-dot difference when the squares contain more dots.

The same principle holds for other kinds of sensations. A small difference in intensity is easier to detect for low-wattage bulbs than for high-wattage bulbs. A small change in loudness is easier to detect for a whisper than a shout. This diminishing effect of increased intensity is called Fechner's law and is illustrated by the logarithmic function in Figure 3.1. The x-axis shows physical intensity and the y-axis shows perceived intensity. Changes in physical intensity produce larger perceptual effects at low levels than at high levels. Our subjective experiences correspond to the log ruler in which small intensities are farther apart (more discriminable) than are large intensities.

Perception often rallies to support cognition but in this case, it works against it. Children will not be good estimators of physical values if their mental number line corresponds to the log ruler. They need a mental number line that has equally spaced intervals such as a tape measure, a scale, or a thermometer. We will begin by looking at how children acquire such a number line and then learn to apply it to a variety of estimation tasks. Then we will look at arithmetic problems that require either estimating temperature after mixing water or estimating concentration after mixing acid. Mixture problems require learning interpolation skills that build on the concept of a mental number line.

$$y = \frac{1}{0.0069}\text{Ln}(x)$$

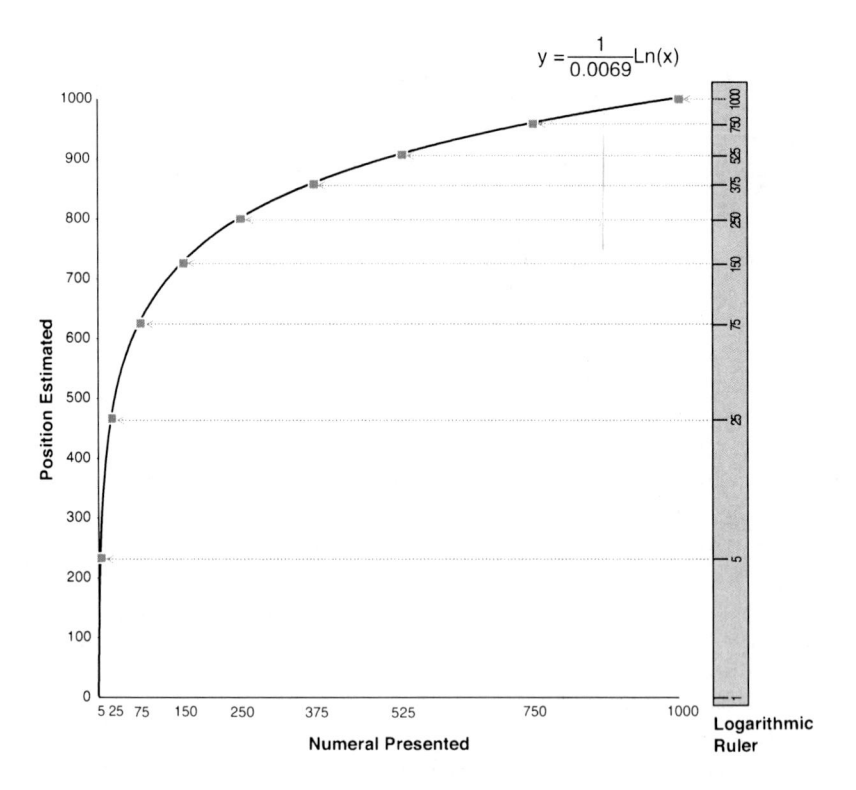

FIGURE 3.1 Predicted location of positions on a number line if children's estimates were based a log rule model. (From "The Development of Numerical Estimation: Evidence of Multiple Representations of Numerical Quantity," by R. S. Siegler and J. E. Opfer, 2003, *Psychological Science, 14*, 237–243. Copyright 2003 by Blackwell. Reprinted with permission.)

THE MENTAL NUMBER LINE

Robert Siegler and John Opfer (2003) at Carnegie Mellon University designed a straightforward procedure for measuring a mental number line. They showed children a blank line that had a 0 at the left end and the number 10, 100, or 1000 at the right end. They then gave the child a number and asked him or her to mark on the line where the number was located. The x-axis in Figure 3.1 shows the numbers he presented for the 0 to 1000 line and the graph shows predictions based on Fechner's logarithmic function. If children could accurately locate numbers on the number line, the graph would be a straight line (or linear function) instead of a curve.

Siegler and Opfer (2003) found that the logarithmic function best fit the estimates of 91% of the second graders, 44% of fourth graders, 28% of sixth graders, and 3% of adults. A linear function best fit the estimates of 97% of the adults, 72% of the sixth graders, 38% of the fourth graders, and 9% of the second graders. The second and fourth graders did much better when they estimated locations on the 0 to 100 line, but many could not generalize their knowledge to the 0 to 1000 line.

Siegler and Ramani (2006) studied even younger children's ability to locate numbers on the 0 to 10 line. This was a difficult task for 4-year-olds from low income backgrounds, although children from middle-income backgrounds were quite accurate. Siegler and Ramani hypothesized that the middle-class children benefited from moving tokens on board games that they likely played in their homes. One implication of this hypothesis is that providing similar experiences to low income children should improve their ability to construct number lines. Over a two-week period children from urban Head Start centers played a board game that required moving tokens along squares. The board consisted of either 10 consecutively numbered squares or 10 colored squares. Consistent with the hypothesis, Head Start children who played on numbered boards improved sufficiently on the number-line estimation task that they now equaled the middle-class children. Children who played on colored squares showed no improvement on the estimation task.

Booth and Siegler (2006) studied whether an accurate mental number line helps children estimate the number of dots in a square and draw a line of a specified length. As anticipated from previous research, second and fourth graders differed in the extent to which their placement of numbers on a line fit a linear function. Those children whose estimates were more linear also showed more linearity in their estimation of dots and in their drawing of lines of a specified length. Possessing a mental number line consisting of equally spaced numbers therefore appears to be a key to performing a variety of spatial estimation tasks.

Booth and Siegler (2006) included a fourth task in their research that measured computational estimation. Children had to estimate answers to problems such as 377 + 82 and 639 − 344. These problems appeared to tap a different set of skills than the spatial estimation problems. Booth and Siegler suggested that computational estimation depends on the ability to memorize arithmetic facts and to use specific procedures such as rounding. Statistical support for making a distinction between spatial and computational estimation also comes from a study of college students by Hogan and Brezinski (2003). Although the two studies were conducted independently, they arrived at similar conclusions.

SPATIAL SKILLS

The similarity between the two studies begins with the three types of estimation problems that Hogan and Brezinski (2003) gave their undergraduates at the University of Scranton. The *numerosity* task required estimating how many dots were in an array that was briefly shown for a quarter of a second. The number of dots varied over the 15 trials, but the average number was 40. The *measurement* task included six items on length (How tall is the library on campus?), four items on weight (How many pounds does this mug of water weigh?), two items on volume (About how many pennies will it take to fill this 1-quart container?), and three items on time (About how many seconds does it take to dial all seven digits of a local telephone number?). The *computational* task consisted of 20 computational problems, such as 97 × 26 = __ and 2/19 + 2/17 = __, but did not allow enough time to calculate exact answers. The undergraduates also took two standardized mathematics tests. One measured their ability to perform basic arithmetic operations and the other measured their ability to solve word problems requiring arithmetic or simple algebra.

The authors used a statistical technique called principal components analysis to infer the underlying skills that are required to perform well on these kinds of tasks. The analysis yielded two principle components that can be interpreted as evidence for two separate sets of skills. The first component was prominent for computational estimation and for performance on the two standardized tests, indicating that computational estimation is part of general mathematics ability. The second component was prominent only in the numerosity and measurement tasks. These estimations therefore form a unique skill, separate from computational estimation and general mathematical ability.

Hogan and Brezinski (2003) suggest that this second component is related to general spatial and perceptual ability. They also suggest that the distinction between these two components corresponds to the distinction made by Dehaene (1997) between an ability to work with exact numbers and an ability to approximate. Although computational estimations involve "approximations," they depend on the same abstract, numeric systems that are used to make exact calculations.

One educational implication of the findings, according to Hogan and Brezinski (2003), is that the development of computational estimation closely parallels general mathematical development. Although students need instruction on rounding, the meaning of approximation, and the need to check the reasonableness of estimates, these skills are closely related to the skills required for exact computations. In contrast, competence in measurement estimation does not follow automatically from general mathematical development and needs to be addressed as a separate skill.

Although the Carnegie Mellon and Scranton studies agree that computational estimation requires different skills than numerosity and measurement estimation, there is a difference in emphasis. Booth and Siegler (2006) point to a particular spatial skill—equal spacing along a mental number line—while Hogan and Brezinski (2003) point to more general spatial skills. It is likely both are important, as suggested by research on instructing children to make measurement estimations of length.

MEASUREMENT

Estimating length requires not only equally spaced numbers but also spaces between numbers that match a standard unit of measurement. The mental measurement of length requires both imagining an object of known length and determining how many of those objects, placed next to each other, would fill the distance. A person could estimate the length of a room by imagining how many yardsticks it would take to fill the distance. The problem, however, with using a standard unit of measurement to instruct children is that they might not know the exact size of an inch, a foot, or a yard unless they could relate it to a familiar object. Even adults are told by their dermatologists to pay attention to moles that are larger than a pencil eraser, rather than to pay attention to moles that are larger than 6 mm.

Elana Joram at the University of Northern Iowa hypothesized that third-grade students would show greater improvement in estimation if they used a familiar object, rather than a standard, as a measurement unit (Joram et al., 2005). They were allowed to select their own reference object such as a toy car or piece of bubble gum that measured approximately 1 inch. Training on familiar objects also included instruction on how a 6-inch dollar bill could be used to measure both 9 inches and 12 inches. Although the children made frequent use of familiar objects during the instruction, they initially compared these objects to rulers to ensure that they would make connections to the standard units of measurement (see Figure 3.2). A control group of third graders received similar instruction based only on standard units.

Both groups later estimated the length of objects that ranged from 3 inches to 9 feet. The test items were familiar objects (length of a paintbrush, height of a door) that had not been used during the instruction. Children who trained on standard units to estimate length made estimation errors that were three times greater than children who trained on familiar objects. They also made errors that were twice as great when asked to draw lines that represented an inch and a foot.

Joram and her co-investigators (2005) mention two potential advantages of familiar objects over standard units of measurement. First, familiar objects can

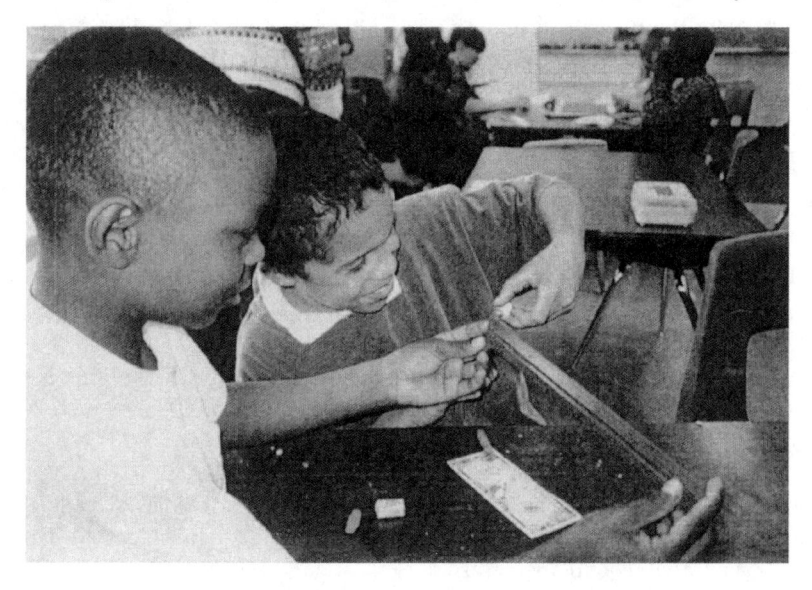

FIGURE 3.2 Measuring the size of reference objects such as an eraser, bubble gum, and dollar bill. Picture provided by Elana Joram.

make measurements more meaningful. This may be particularly true when students have their choice of objects to use as a referent. Children are likely to better remember the length of an inch as a piece of bubble gum than as a mark on a ruler. Their more accurate drawings of lines that are an inch long and a foot long provide empirical support for this claim. Second, using familiar objects can reduce the number of mental alignments that are required. Estimating that an object is 9 inches long should be easier when realizing that it is equal to 1.5 dollar bills than when mentally locating the placement of nine 1-inch units.

MIXTURE PROBLEMS

The concept of a mental number line is also helpful for explaining how people estimate answers to mathematics problems. As an example, estimate an answer to the Temperature Problem.

Temperature Problem

A science teacher mixes 2 cups of 40° water with 3 cups of 60° water. What is the temperature of the mixture?

Two principles that are useful for estimating answers to mixture problems are the range principle and the quantity principle. The *range* principle states that the temperature of a mixture will be between the temperature of the cooler solution (such as 40°) and the temperature of the warmer solution (such as 60°). This principle establishes a mental number line that is anchored at one end by the cooler temperature and at the other end by the warmer temperature. The *quantity* principle states that increasing the quantity of the warmer solution will increase the temperature of the mixture. This principle provides a rule for moving along the number line.

A question raised by this example is "How does the knowledge of principles influence the selection of mathematical procedures for solving a problem?" Are principles used merely to constrain the answers in an estimation task or do they influence calculations in a mathematical task? The answer to the temperature problem can be calculated by using proportional reasoning. The mixture is two-fifths 40° water and three-fifths 60° water so two-fifths of 40° plus three-fifths of 60° is 52°.

Ahl, Moore, and Dixon (1992) investigated this question at the University of Wisconsin by comparing qualitative reasoning based on principles to quantitative reasoning based on calculation. They were influenced by Inhelder and Piaget's (1958) claim that a qualitative grasp of proportions precedes children's ability to manipulate numerical proportions. One implication of such a claim is that asking students to estimate might enhance their subsequent ability to calculate.

Estimation 31

Mathematical Task:

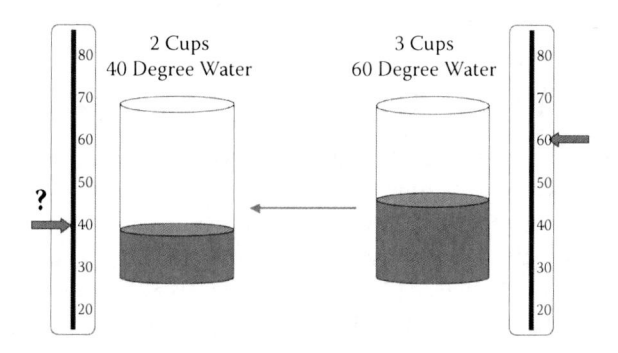

FIGURE 3.3 Example of a quantitative task in which students attempt to calculate the new temperature after adding 3 cups of 60° water to 2 cups of 40° water. (Based on "The Developmental Role of Intuitive Principles in Choosing Mathematical Strategies," by J. A. Dixon and C. F. Moore, 1996, *Developmental Psychology, 32*, 241–253.)

The investigators tested this hypothesis by designing a quantitative and a qualitative version of the temperature-mixture task. An example of a quantitative (mathematical) problem, shown in Figure 3.3, is the example that we have already considered. Students were encouraged to try to solve the problem by using mathematics. Other variations of this problem involved mixing two solutions that contained 1, 2, or 3 cups of water and had temperatures of 20°, 40°, 60°, or 80°.

Figure 3.4 shows the qualitative (estimation) version of the task. The three amounts were labeled "small," "medium," and "large" and were illustrated by the amount of water in the containers. The four temperatures, labeled "cold," "cool,"

Estimation Task:

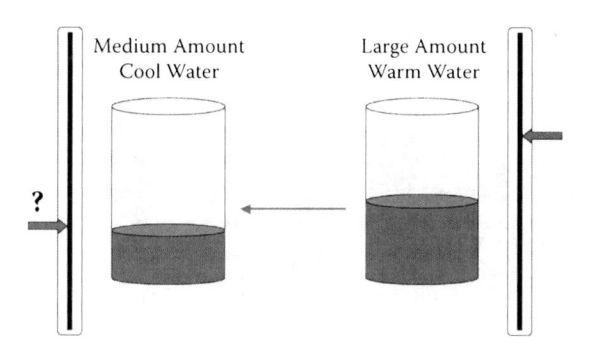

FIGURE 3.4 Example of a qualitative task in which students estimate the new temperature after adding a large amount of warm water to a medium amount of cool water. (Based on "The Developmental Role of Intuitive Principles in Choosing Mathematical Strategies," by J. A. Dixon and C. F. Moore, 1996, *Developmental Psychology, 32*, 241–253.)

"warm," and "hot," were depicted by the height of the arrows on unmarked thermometers. Students responded by adjusting the height of the left arrow to estimate the temperature of the mixture if the water in the right container were added to the left container.

College students gave fairly accurate answers on both the estimation and mathematical tasks. However, fifth and eighth graders did better on the mathematical task when they first performed the estimation task. Of the 65 students who did the mathematical task first, 23 simply added the temperatures of the two solutions, compared to only 9 of 71 students who did the estimation task before the mathematical task. This finding suggests that the intuitive principles used to constrain estimates can also help the students select better mathematical procedures.

A follow-up study by Dixon and Moore (1996) confirmed this hypothesis. In their experiment all of the students (second, fifth, eighth, and eleventh grade, and college) performed the estimation task before performing the mathematical task so the experimenters could measure how well students understood the range and quantity principles. Figure 3.5 shows how the understanding of these principles increases with grade level. Understanding a principle was found to be a necessary, but not a sufficient, condition for selecting an appropriate strategy for the mathematical task. Very few students used a math strategy that was consistent with a principle that they did not understand. However, some students occasionally used math strategies that violated principles that they did understand, perhaps because they could not think of an alternative strategy.

A particularly interesting finding was that students' answers were more accurate in the estimation task than in the mathematical task. The experimenters

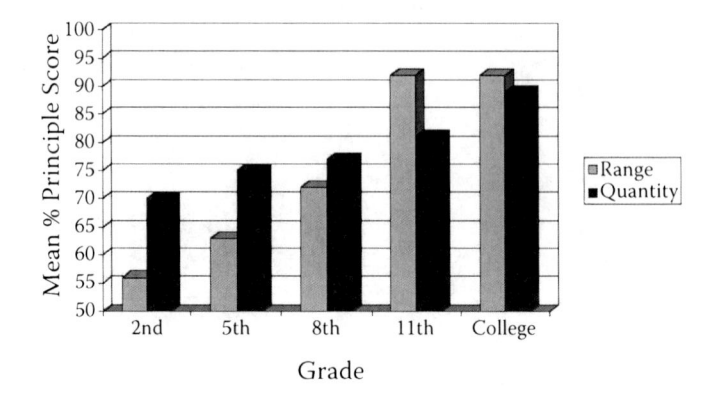

FIGURE 3.5 Development of the range and quantity principles. (Based on "The Developmental Role of Intuitive Principles in Choosing Mathematical Strategies," by J. A. Dixon and C. F. Moore, 1996, *Developmental Psychology, 32,* 241–253.)

Estimation 33

converted students' estimates to numbers by superimposing the numerical thermometers in Figure 3.3 over the unmarked thermometers in Figure 3.4. The average absolute error was 4° in the estimation task and 16° in the mathematical task. Although the estimates were fairly close to the correct answer, the mathematical strategies often produced answers that were not very accurate. In this particular task, intuitive understanding based on principles was more reliable than the use of mathematical strategies.

A question raised by these findings is whether this intuitive understanding requires thinking visually. I would argue that this particular task requires mapping our sensory experiences about temperature onto a visual line—the unmarked thermometer in Figure 3.4. The perceptual mapping likely involves visual thinking.

INTERPOLATION

Now test your estimation powers on a mixture task that may be less familiar to many of you. The Acid Problem specifies mixing two solutions that differ in concentration of acid.

Acid Problem

Estimate the concentration of a 10-pint mixture when:

1. 1 pint of a 10% concentration is mixed with 9 pints of a 40% concentration.
2. 3 pints of a 10% concentration is mixed with 7 pints of a 40% concentration.
3. 5 pints of a 10% concentration is mixed with 5 pints of a 40% concentration.
4. 7 pints of a 10% concentration is mixed with 3 pints of a 40% concentration.
5. 9 pints of a 10% concentration is mixed with 1 pint of a 40% concentration.

If we know that the range and quantity principles apply to mixing acid, we should know that the concentration of the mixture varies from 10% to 40% and decreases as the proportion of the 40% solution decreases. However, we can make estimates that are more precise (correct) by using linear interpolation. We can determine *how much* the concentration decreases for each 1-pint change by dividing the range by the number of pints: (40% − 10%)/10 = 3%. Each 1-pint change lowers the concentration by 3%. The correct answers to the questions are (1) 37%, (2) 31%, (3) 25%, (4) 19%, and (5) 13%. Notice that linear interpolation is consistent with establishing a mental number line that is an important theoretical concept in much of the research on numerical reasoning.

What would happen if the problem involved creating 10 pints of water by mixing water that had a temperature of 10°C with water that had a temperature of 40°C? Reed and Evans (1987) attempted to answer this question by comparing the ability of undergraduates at Florida Atlantic University to make accurate estimates on the acid task with their ability to make accurate estimates on the temperature task. Students in the temperature group saw mixtures with the same numbers as students in the acid group but the units in the principles and test questions were labeled °C rather than % *acid*.

The results showed that students' estimates in the acid task improved after they studied the principles of range, quantitiy, and linearity, but were still much worse than the students' initial estimates in the temperature task. Students came to the experiment with a good understanding of how these principles applied to temperature. We gain an understanding of these principles throughout our experiences.

At the end of Chapter 1, I mentioned that spatial reasoning does not necessarily require visual reasoning. Interpolation involves spatial reasoning but it can be done by following verbal instructions (mathematical procedures). Does interpolation therefore involve visual thinking? My belief is that the acquisition of all the estimation skills reviewed in the chapter depends on a visual representation of the mental number line discussed in Chapter 2. I hope my review stimulates further research on this topic to evaluate my belief.

SUMMARY

This chapter provides evidence for the importance of spatial skills in estimating answers to a variety of measurement tasks. Siegler's research (Siegler & Opfer, 2003; Siegler & Ramani, 2006) demonstrates that children's initial number line is logarithmic and young children have to learn to equally space numbers. Hogan and Brezinski (2003) point to the importance of spatial skills in estimation tasks that do not involve computational estimation. They argue that mathematics instruction should include exercises for developing these skills. Joram's research (Joram et al., 2005) shows that the substitution of familiar objects for standard units is effective in improving children's measurement skills. The use of interpolation in solving mixture problems also builds on the concept of a mental number line. The range and quantity principles that support interpolation are gradually acquired by children, but can eventually be used to guide selection of appropriate mathematical procedures as demonstrated by Dixon and Moore (1996) and Ahl et al. (1992). My own research (Reed, 1999, pp. 174–180) has shown that the much better estimates for temperature than for concentration are caused by greater use of the interpolation strategy for temperature mixtures.

4 Spatial Metaphors

The mental number line provides a foundation for numerical reasoning and spatial metaphors provide a foundation for understanding abstract ideas by mapping them onto concrete ideas. Remember Truman Capote's (1965) metaphor from Chapter 1: "Like the waters of the river, like the motorists on the highway, and like the yellow trains streaking down the Santa Fe tracks, drama, in the shape of exceptional happenings, had never stopped there." And Kaye Gibbons' (2006) description of Ellen Foster's mother: "The hole was emptier than holes with merely nothing in them." Creative writers breathe life into abstract ideas through their use of metaphor.

However, according to Lakoff and Johnson (1980) in their book *Metaphors We Live By*, metaphors are not limited only to writers' poetic imaginations but form the basis for many of our concepts. They argue that our conceptual system, which influences the way we think and act, is fundamentally metaphorical in design. Lakoff and Johnson met in January 1979 and broke down the walls that separate academic departments. Lakoff, a linguistics professor at the University of California, Berkeley, and Johnson, a philosophy professor at the University of Oregon, shared an interest in metaphor that was driven by their different backgrounds in linguistics and philosophy. Their discussions revealed that both fields had largely ignored the metaphorical nature of thought.

Spatial metaphors, in particular, provide a foundation for many concepts. Table 4.1 shows ones that depend on the distinction between *up* and *down*. By the time Lakoff and Johnson had finished writing their book, they had reached a number of conclusions:

- Most of our fundamental concepts are organized in terms of one or more spatial metaphors. Table 4.1 lists a variety of examples.
- There is a consistency in these metaphors. Positive concepts such as consciousness, health, control, good, and virtue are up. Negative concepts such as unconscious, sickness, subordination, bad, and depravity are down.
- Spatial metaphors are rooted in physical and cultural experiences. We usually are standing or sitting up when conscious and healthy, and lying down when unconscious or sick.
- Spatial metaphors provide a foundation for more advanced thinking in academic fields. Cognitive psychology and mathematics are examples.

TABLE 4.1
Concepts Based on the Up–Down Metaphor

Conscious is Up	Unconscious is Down
Wake *up*.	She *fell* asleep.
He *rises* early.	He's *under* hypnosis.

Health is Up	Sickness is Down
She's at the *peak* of health.	She came *down* with the flu.
Lazarus *rose* from the dead.	He *dropped* dead.

Having Control is Up	Being Controlled is Down
He has control *over* her.	She is *under* his control.
She is on *top* of the situation.	He is *low* man on the totem pole.

Good is Up	Bad is Down
Things are looking *up*.	Things are at an all-time *low*.
He does *high*-quality work.	It's been *downhill* ever since.

Virtue is Up	Depravity is Down
He is *high*-minded.	That would be *beneath* him.
She is an *upstanding* citizen.	She wouldn't *stoop* to that.

Based on *Metaphors We Live By*, by G. Lakoff and M. Johnson, 1980, Chicago: University of Chicago Press.

APPLICATIONS TO COGNITIVE PSYCHOLOGY

I also met George Lakoff in 1979 when I took a sabbatical at Berkeley to write a textbook on cognitive psychology. In 1979, there were very few cognitive psychology textbooks. I had been using a 1967 book by Ulric Neisser that was about cognitive psychology but was not written as a textbook. Neisser began by defining cognitive psychology as referring "to all processes by which the sensory input is transformed, reduced, elaborated, stored, recovered, and used" (Neisser, 1967, p. 4).

The transformation of the sensory input can be depicted by the flowchart shown in Figure 4.1, which I included in the first chapter of my textbook (Reed, 1982). The *sensory store* holds sensory information for a brief time after the stimulus disappears. If you are in a dark room and move a flashlight in a circular motion, you will see a complete circle if you move fast enough to maintain all the points of light from the circular path in your sensory store. The *filter* represents attention and enables you to block out some of the sensory input so you can focus on aspects of your environment that are more relevant. The *pattern recognition* stage is where you convert sensations such as sounds or colors into recognizable patterns. The *selection* stage is where you select information into *short-term memory* (which holds a limited amount of information for 20 to 30 seconds), from which some of it may enter your

Spatial Metaphors

[Diagram: Input → Sensory store ↔ Filter ↔ Pattern recognition ↔ Selection ↔ Short-term memory ↔ Long-term memory; Response arrow above Selection]

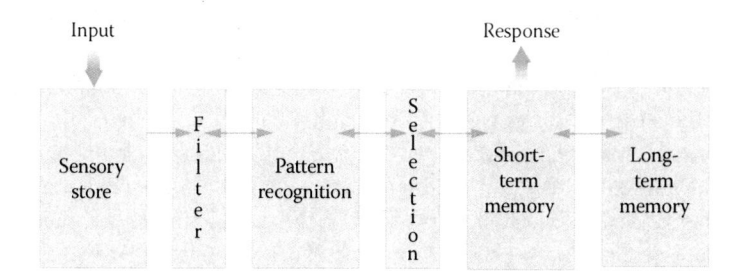

FIGURE 4.1 Information processing stages showing the flow of information between the sensory store and long-term memory. From *Cognition: Theory and Applications,* by S. K. Reed, 1982, Monterey, CA: Brooks/Cole. Copyright 1982 by Brooks/Cole. Reprinted with permission.

long-term memory (which holds an unlimited amount of information from minutes to years).

Lakoff volunteered to read drafts of my first three chapters to show me how metaphors underlie many of the assumptions about cognitive psychology. He pointed out that a flow chart, such as the one shown in Figure 4.1, is built on a path metaphor because it shows the flow of information through various stages between the sensory store and long-term memory. Lakoff also pointed out that the container metaphor underlies our understanding of the sensory store, short-term memory, and long-term memory. Each of these stores can hold information, and much work in cognitive psychology is directed at understanding how these stages operate.

You may have noticed that the arrows between the stages have two points, indicating that information can flow in both directions. The flow from the sensory store to long-term memory makes the obvious point that we need sensory information to recognize patterns. The flow from long-term memory to the sensory store makes the less obvious point that pattern recognition also depends on our expectations. It would probably take you more time to recognize a refrigerator in your living room than a refrigerator in your kitchen. The flow from the sensory store to long-term memory is called *bottom–up* processing. The flow in the reverse direction is called *top–down* processing. I don't know the origin of these terms but assume that the diagram in Figure 4.1 may have initially been drawn vertically rather than horizontally. Alternatively, perhaps to be consistent with the concepts in Table 4.1, someone decided that bottom–up processing is inferior to top–down processing. In either case, spatial metaphors return to influence our thinking.

APPLICATIONS TO MATHEMATICS

Perhaps more than any other domain, the field of mathematics has aroused diverse emotions. To some, such as Lakoff and Nuñez (2000), mathematics is "one of the most profound and beautiful endeavors of the imagination that human beings have ever engaged in" (p. 5). To others, mathematics is a frightening subject that should be avoided at all costs.

Mathematics can be made less intimidating according to Lakoff and Nuñez (2000) if instruction emphasized its metaphorical basis. In their book *Where Mathematics Comes From: How the Embodied Mind Brings Mathematics into Being,* they argue for the important role of metaphor in creating and understanding mathematics. Their argument is based on three assumptions that build on the ideas expressed in *Metaphors We Live By* (Lakoff & Johnson, 1980):

1. The detailed nature of our bodies, brains, and everyday functioning in the world shapes our concepts and reasoning, including mathematical concepts and mathematical reasoning.
2. Most thought is unconscious so we cannot look directly at our conceptual systems and low-level thought processes.
3. Human beings typically conceptualize abstract concepts in concrete terms derived from the sensory-motor system.

The first assumption is that mathematics is not an abstract system that is unrelated to our everyday experiences. For example, we can think of arithmetic as object collection in which addition is putting collections together to form larger collections and subtraction is taking smaller collections from larger collections. Even the understanding of such abstract concepts as the continuity of functions depends on our ability to relate these concepts to our everyday experiences. The second assumption is that the metaphors that underlie thinking are often unconscious. You probably have heard of the expression "time is money" and can recall many experiences in which it is literally true. But you may not have thought about the many times you unconsciously use the metaphor in language and thought. You can both *save* time and *spend* time. You adjust your clocks for daylight-*savings* time even though you do not literally save time when you make the adjustment. You may know people who have lived on *borrowed* time. The third assumption is that our sensory-motor system helps us understand abstract concepts. Even mathematicians may refer to the sensory-motor system, as did Euler when he described a continuous function as a curve described by freely leading the hand.

Another example of understanding the abstract by mapping it onto the concrete is to view categories as containers. The categories-as-containers metaphor enables logical inference to be represented as spatial inference in Lakoff and Nuñez's (2000) analysis. A category becomes a bounded region in space, and members of the category are objects inside the bounded region. The use of Venn diagrams to represent classes and sets makes use of this metaphor, as shown in Figure 4.2. Notice in Figure 4.2a, an object x placed in container A is also in container B if container A is inside container B. Object y is outside both containers. The Venn diagram in Figure 4.2b is an abstract representation based on the container metaphor. The Missing Dollar Problem illustrates how this diagram is useful.

Spatial Metaphors

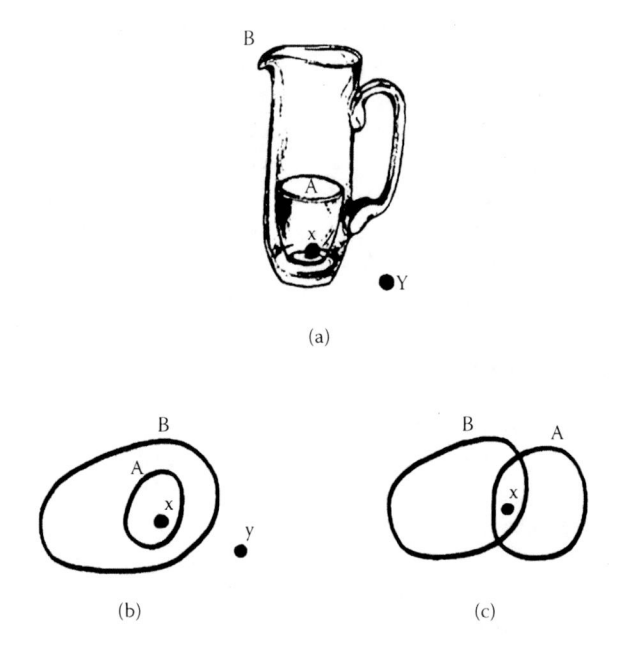

FIGURE 4.2 A container (a) that corresponds to the Venn diagram in (b) but does not correspond to the Venn diagram in (c). From *Where Mathematics Comes From: How the Embodied Mind Brings Mathematics into Being,* by G. Lakoff and R. E. Nuñez, 2000, New York: Basic Books. Copyright 2000 by Basic Books. Reprinted with permission.

The Missing Dollar Problem

Some years ago, three students rented a motel room. They each paid $10 for the $30 room. The manager then realized that on weekdays the room rented for $25. He gave $5 to the bellhop with instructions to return it to the students. The bellhop decided that because it would be difficult for three people to divide $5, he would keep $2 as his tip. Each student therefore paid $9 for the room, for a total of $27. The bellhop kept the remaining $2. What happened to the missing dollar?

The solution of the problem requires listening to Deep Throat's advice during the Watergate scandal. The Venn diagram in Figure 4.2b helps us "follow the money." Let container B hold the $27 that the students paid for the room. Part of the $27 (in container A) is the $2 tip kept by the bellhop. The remainder of the $27 is the cost of the room. The rest of the money (represented by y) is the $3 returned to the students. The missing dollar was created by shifting attention away from the refunded $3 to the $2 dollar tip, which should not be added to the $27 because it is included in the $27.

An advantage of Venn diagrams is that they allow for greater flexibility than physical containers. Although it is easy to imagine one container in another, it is less easy to imagine two overlapping containers such as depicted by the Venn diagram in Figure 4.2c. In this example, we might let category A be property owned by one couple and category B be property owned by their neighbors. The overlap of the two categories represents a jointly owned fence that runs along the property line. Although you probably have not experienced overlapping containers, you can still make use of the general idea that a category is a bounded region in space and members of the category are objects inside the bounded region.

GROUNDING METAPHORS

Lakoff and Nuñez (2000) distinguish between two fundamental kinds of metaphor: grounding metaphors and linking metaphors. *Grounding metaphors* relate ideas to everyday experiences. The container metaphor is an example of a grounding metaphor because it enables us to think about a Venn diagram as a container of objects or about memory as a container of our experiences. The path metaphor is another example of a grounding metaphor because it enables us to think about ideas as movement along a path.

Lakoff and Nuñez (2000) have expanded the path metaphor into a more elaborate structure that they call the source–path–goal schema, depicted in Figure 4.3. This structure represents motion as a trajectory moving along a path from a source to a goal. One example of the source–path–goal schema is the concept of fictive

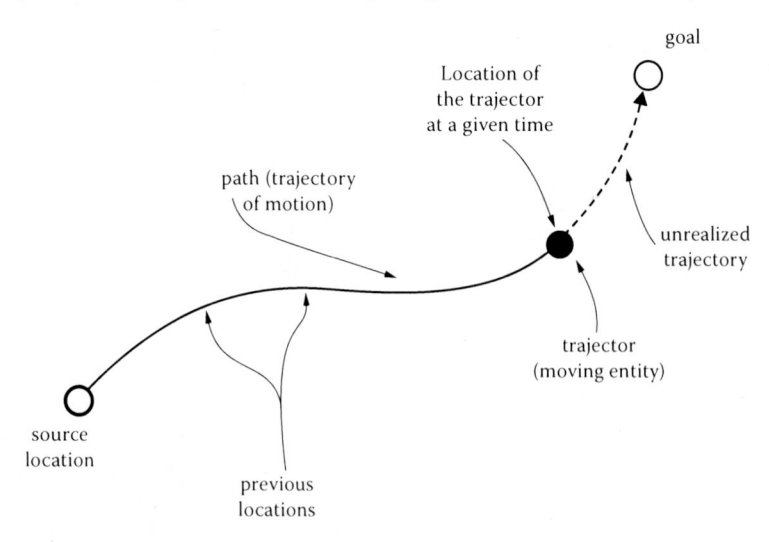

FIGURE 4.3 The source–path–goal schema showing the trajectory of motion along a path from the source to the goal. (From *Where Mathematics Comes From: How the Embodied Mind Brings Mathematics into Being*, by G. Lakoff and R. E. Nuñez, 2000, New York: Basic Books. Copyright 2000 by Basic Books. Reprinted with permission.)

motion, which states that a line represents the motion of a traveler tracing that line. Fictive motion is used when people make statements such as "The road *goes through* the desert" or "The fence *runs along* the property line." Phrases such as *goes through* and *runs along* suggest motion, although highways and fences do not literally move.

We saw evidence in Chapter 1 for the perceptual symbols hypothesis that understanding verbal statements depends on cognitive simulation. For instance, people more quickly verified that "nail" was mentioned in a sentence when the orientation of a nail in a picture matched the implied orientation of the nail in the sentence. Although it is easy to simulate the act of pounding, how can anyone simulate figurative motion such as a fence running?

Surprisingly, Matlock (2004) has found evidence that fictive motion encourages us to imagine travel. Undergraduates at the University of California, Santa Cruz read about either a short-distance scenario in which it takes Maria only 20 minutes to drive on Road 49 across the desert or a long-distance scenario in which it takes Maria 7 hours to drive on Road 49 across the desert. Matlock hypothesized that people would construct a simulation of the scenario as they read the sentences. After reading the scenario they had to indicate whether a test sentence (Road 49 crosses the desert) was related to the story. People took much longer to verify the test sentence in the long-distance scenario than in the short-distance scenario, suggesting that they thought about travel along the road when they made their verification. The long- versus short-distance scenarios did not influence verification times for test sentences that did not involve fictive motion (Road 49 is located in the desert).

If we can mentally simulate fictive motion, it should not be surprising that we can simulate movement along a number line. Lakoff and Nuñez (1997) proposed that imagining this movement enables us to represent arithmetic as motion. The size of a number is then the length of the trajectory from the origin to the location on the line. Addition is taking steps a given distance to the right (or forward) and subtraction is taking steps a given distance to the left (or backward).

Cemen (1993) described how this method could be physically acted out to help children understand arithmetic operations. A teacher could place a number line of positive and negative numbers on the floor. The distance between numbers should be about the size of a child's footstep. The rules are that (1) positive numbers require stepping forward, (2) negative numbers require stepping backward, and (3) subtraction requires reversing direction. Adding 3 to 5 requires moving forward 5 steps and then taking 3 additional steps forward to 8. Subtracting 5 from 3 requires taking 3 steps forward, turning around for the subtraction sign, and moving forward (now toward the smaller numbers) 5 more steps to land on –2. Even the tricky problem 4 – (–3) can be solved by this method. The child moves forward 4 steps and then turns around (reversing direction) when encountering the subtraction sign. The child then takes 3 steps backward for the –3 to finish at 7.

Notice that the concept of the number line continues to evolve. We first encountered an implicit number line as an explanation of why it takes longer to decide that 66 (rather than 90) is greater than 65. We again encountered this concept in the research of Siegler and Opfer (2003) on children's gradual acquisition of a number line with equally spaced units. Mentally aligning familiar units was a key in Joram, Gabriele, Bertheau, Gelman, and Subrahmanyam's (2005) research in helping students better estimate the length of objects. Finally, we see how physically acting out addition and subtraction by taking steps along a number line may help students learn arithmetic.

SEARCHING SEARCH SPACES

The source–path–goal schema provides a metaphor for thinking about not only the specific case of arithmetic as movement along a number line, but for thinking about all kinds of problems as movement toward a goal. Problems consist of a starting point and a goal, and we hope progress toward the goal. For example, a starting point could be a person's beginning weight, a goal could be a person's desired weight, and progress could be measured by the difference between the current and desired weight.

Newell and Simon (1972) at Carnegie Mellon University conceived of problem solving as searching through a search space. The search space consists of an initial state, a goal state, and all the intermediate states between the initial and goal states. Early in my career, I became interested in how search spaces influence problem solving and began to study a problem that has about the right size search space; big enough to be interesting but small enough that people could solve it. It is a more complex version of the standard Missionaries–Cannibals Problem.

Five Missionaries–Cannibals Problem

Five missionaries and five cannibals who have to cross a river find a boat, but the boat is so small that it can hold no more than three persons. If the cannibals outnumber the missionaries on either bank of the river or in the boat at any time, the missionaries will be eaten. Find the simplest schedule of crossings that will allow everyone to cross safely. At least one person must be in the boat at each crossing.

As with other problems in this book, the discussion will be more relevant if you try to solve the problem. As an incentive, see if you can solve it in fewer than the average of 30 moves that occurred in my research.

The average of 30 moves occurred for students at Case Western Reserve University who were not given a hint. Another group of students in the subgoal group was told that solving the problem required reaching a problem state in which there were three cannibals across the river by themselves and without the boat. Students in the subgoal group took an average of 20 moves to solve the problem.

Spatial Metaphors

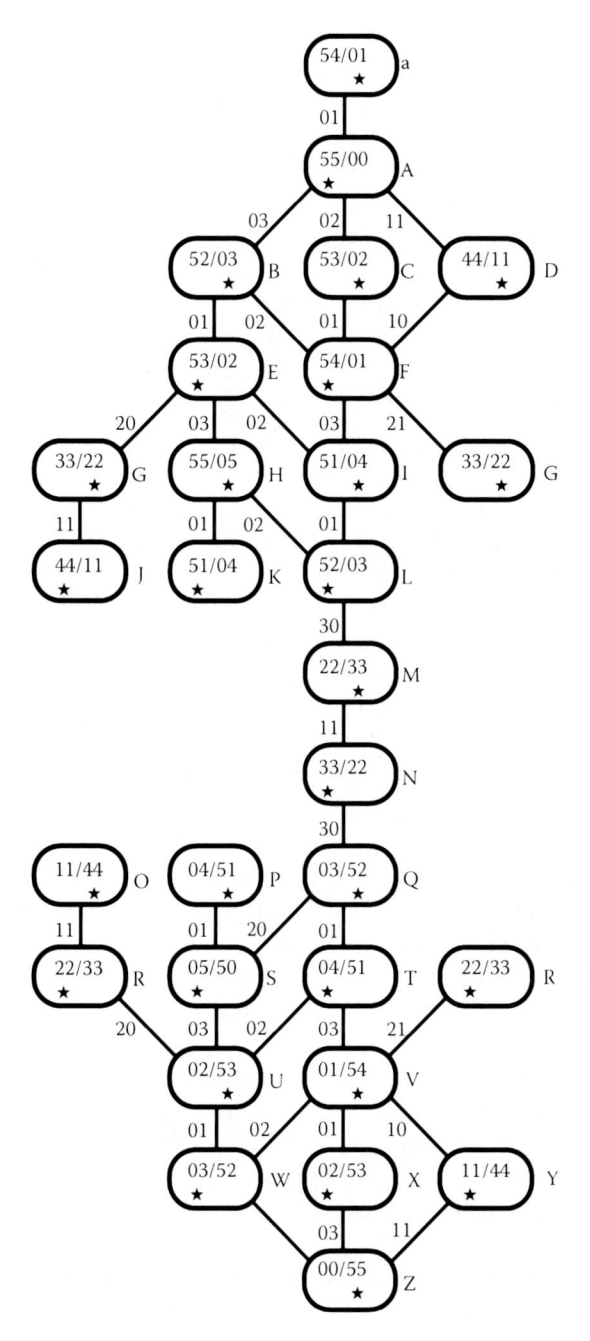

FIGURE 4.4 A search space for the Missionaries–Cannibals Problem. (From "Modeling Strategy Shifts in a Problem-Solving Task," by H. A. Simon and S. K. Reed, 1976, *Cognitive Psychology, 8*, 86–97. Copyright 1976 by Elsevier. Reprinted with permission.)

Studying problem solving is much easier if we understand the search spaces in which people search for solutions. Figure 4.4 shows the search space (all the permissible moves and problem states) for the Missionaries–Cannibals Problem. The numbers next to the lines show the number of missionaries (first digit) and the number of cannibals (second digit) moved in the boat. The numbers in the ovals show the number of missionaries and cannibals on the near bank followed by the number of missionaries and cannibals on the far bank. The asterisk shows the location of the boat. We can learn from many characteristics of the problem by studying the search space. For example, the problem can be solved in 11 moves, entering the "blind alley" (state J) requires backing up, and it is necessary to go through the subgoal state (state L) in order to reach the goal state (state Z).

To try to understand why the subgoal was so effective, Simon and Reed (1976) developed a simulation model of how students in the two groups explored the search space. The goal of the model was to predict, for each of the possible legal moves, the average number of times students in each group would make that particular move. We initially thought students would follow a *means–end strategy* in which they would move as many people across the river as possible (typically three) and bring back as few as possible (typically one). A model based on the means–end strategy was fairly successful in predicting their choices, but some of their moves did not follow this strategy.

Violations of the means–end strategy occurred when problem solvers followed a *balance strategy* that attempts to create equal numbers of missionaries and cannibals on each side of the river. For example, the most frequent initial move was to take across one missionary and one cannibal even though it is possible to take across three cannibals. The balance strategy makes it easy to avoid illegal moves because cannibals will never outnumber missionaries as long as the numbers of missionaries and cannibals are equal on both sides. The trouble with the balance strategy is that it tends to lead people away from the solution path and toward the blind alley because both states G and J are balanced problem states.

After analyzing the moves made by both groups, we developed a simulation model in which the subgoal group switched sooner from the balance strategy to the more effective means–end strategy. The fact that the subgoal—three cannibals and no missionaries—is an unbalanced problem state also makes it intuitively likely that students in the subgoal group would not persist in following the balance strategy.

We have seen that the source–path–goal schema is helpful for thinking about a specific problem (arithmetic) as movement along a number line and, more generally, as movement toward a goal. Both interpretations are made possible by the same grounding metaphor based on our experiences in moving along a path. In the next section, Lakoff and Nuñez (2000) introduce us to a second kind of metaphor called a linking metaphor.

LINKING METAPHORS

While grounding metaphors allow us to ground our understanding in familiar experiences, *linking metaphors* allow us to link one domain of knowledge with another domain of knowledge. For instance, they enable us to better understand abstract mathematical ideas such as functions by graphing them.

A nice example is the 106-page footnote that René Descarte attached to his book *Discourse on Method* published in 1637 (translated into English by Lafleuer [1960]). The footnote, known as La Geometrie, proposed that a pair of numbers could be plotted on a two-dimensional surface. This insight helped launch graph paper, analytic geometry, and a linking metaphor between numbers and space.

There are, of course, thousands of examples in which plotting numbers is helpful, but let me focus on only one—multidimensional scaling. There are many occasions when it is helpful to measure the perceived similarity of two objects or concepts by representing similarities as distances between points in a (Cartesian) coordinate space. Figure 4.5 shows a two-dimensional scaling solution based on similarity ratings of birds that were collected by Rips, Shoben, and Smith (1973) at Stanford. The distance between the points represents similarities. For instance, goose and duck were rated as very similar but goose and eagle were not.

Now you might ask why go to the trouble of feeding these similarity ratings into a multidimensional scaling program. Why not use similarity ratings to

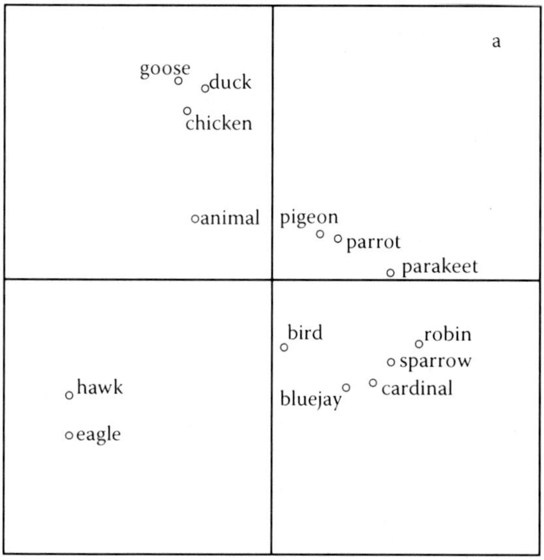

FIGURE 4.5 A multidimensional scaling solution in which similar concepts are nearer each other. (From "Semantic Distance and the Verification of Semantic Relations," by L. J. Rips, E. J. Shoben, and E. E. Smith, 1973, *Journal of Verbal Learning and Verbal Behavior, 12*, 1–20. Copyright 1973 by Elsevier. Reprinted with permission.)

represent similarities? One answer is that scaled distances between concepts are more accurate measures of similarity than the ratings themselves. The reason is that the location of each bird in the multidimensional space is constrained by *all* the similarity ratings of that bird. The distance between *robin* and *parrot* is determined not only by the judged similarity of this pair but also by the judged similarity of all the pairs that involve robin or parrot.

A second benefit of multidimensional scaling is that it is sometimes possible to interpret which features influence the similarity ratings by examining the axes. Look at the location of the points along the vertical axis and see if you can label the dimension to account for the ordering of birds such as goose and duck on the top and hawk and eagle on the bottom. Now try to do the same for the horizontal axis as the birds vary from hawk and eagle on the left to robin and sparrow on the right. There are some exceptions but the vertical dimension might be labeled "predacity" in which goose, duck, and chickens are (mostly our) prey and hawks and eagles are predators. The horizontal dimension might be labeled "size" in which the larger birds appear at the left and the smaller birds at the right.

A third benefit, which is a general advantage of plotting data, is that data plots reveal patterns that are difficult to find by looking at numbers in tables. We can immediately see, for example, that *goose*, *duck*, and *chicken* are perceived as being very similar to each other and very different from other birds. A clever addition to this particular experiment is that people judged not only the similarity of birds to each other, but also the similarity of each bird to their concept of a bird and to their concept of an animal. We can now see which birds are judged more typical by looking at their distances from the point labeled "bird" in the multidimensional space.

The primary focus of this book is how we all use visual thinking in our daily lives. However, a secondary theme is how cognitive scientists represent people's thinking. This chapter illustrates how their models and theories are often based on spatial representations.

SUMMARY

In their book, *Metaphors We Live By*, Lakoff and Johnson (1980) argued that metaphors form the basis for conceptual thinking. Many of these metaphors, such as the up–down classification of concepts, are spatial. Even abstract domains such as mathematics rely on metaphors for understanding. Venn diagrams, for example, provide a container metaphor of how people represent category relations. Psychologists have used spatial representations extensively to formulate theories. The flow chart in Figure 4.1 represents a model of how people process information at various stages between the sensory input and long-term memory. The search space in Figure 4.4 enables us to understand better how people solve problems. Moreover, the multidimensional scaling solution in Figure 4.5 shows how people represent similarity within categories. Working behind the scenes in all of these visual representations are spatial metaphors as simple as paths and containers.

5 Producing Images

This chapter and the next discuss the benefits of visual imagery. You might think we have already been discussing visual images. I have written about babies' and animals' ability to remember whether there were one, two, or three hidden objects. I mentioned babies' memory for faces. I have written about the role of perceptual simulation in understanding language. I have shown how spatial metaphor influences our thinking and the theories of cognitive psychologists as they try to model our thinking. It is likely that visual imagery plays an important role in many of these activities.

However, this research has been on the periphery of the study of visual imagery. One of the reasons is that some of the cognitive processes that support these activities occur unconsciously and therefore we are not aware of them. Remember that one of Lakoff and Nuñez's (2000) central assumptions was that most thought is unconscious. I disagree that most thought is unconscious, but some of it is, particularly when it operates at the level of metaphor.

The perceptual symbols systems proposed by Barsalou (1999) can also operate unconsciously: "Most importantly, the basic definition of perceptual symbols resides at the neural level: unconscious neural representations—not conscious mental images—constitute the core content of perceptual symbols" (p. 583). So, for example, the mental simulation of travel when comprehending the sentence "The road crosses the valley" might not be a conscious experience. One of the challenges for perceptual symbols theory is to sort out which experiences are conscious and which are not. In the meantime, we can examine research in which visual images are consciously used to improve memory and spatial reasoning.

The conscious production of mental representations, such as images, to support goal-directed cognition was a central interest of American functionalism. The functionalist approach flourished in the late 19th century and early 20th century through the leadership of William James and John Dewey. However, the functionalists' study of imagery did not survive John Watson's 1924 book *Behaviorism* that attacked all mental constructs. Although functionalists' research methods were sometimes inadequate, many of their claims continue to influence modern thinking according to Pani (1996). The more advanced research methods of the cognitive sciences resurrected visual images to demonstrate that they are too important in cognitive functioning to keep hidden forever.

REMEMBERING VISUAL IMAGES

Visual imagery began to regain its respect through the work of Allan Paivio at the University of Western Ontario in the late 1960s. After an extensive series of studies, Paivio (1969) proposed that there are two major ways a person can elaborate on material to improve learning. One form of elaboration emphasizes verbal associations. A word like *poetry* may result in many associations that could help you distinguish it from other words. You might think of different styles of poetry, particular poems, or experiences in an English class. The other form of elaboration is creation of a visual image to represent a word. If I asked you to remember the word *juggler*, you might form an image of a person juggling balls. If I asked you to remember the word *truth*, however, you would probably have difficulty forming an image. It is easy to form an image to represent a concrete object such as juggler but difficult to form an image for an abstract concept such as truth. At the concrete end of the continuum are pictures, because the picture itself can be remembered as a visual image and the person does not have to create one. Pictures often result in better memory than do concrete words, which usually result in better memory than abstract words.

If visual images and verbal associations are the two major forms of elaboration, is one more effective than the other? To answer this question, we have to know how easy it is to form either an image or a verbal association of a word. The imagery potential of words is typically measured by asking people to rate how easy it is to form an image for a given word. As we might expect, concrete words are rated high on imagery and abstract words are rated low. The association value of a word is typically measured by asking people to give as many associations as they can over a 1-minute interval. Paivio and his colleagues found that the imagery potential of words is a more reliable predictor of learning than the association potential of words (Paivio, 1969). High-imagery words are easier to learn than low-imagery words, but high-association words are not necessarily easier to learn than low-association words.

The reason images are effective, according to Paivio (1986), is that an image provides a second memory code that is independent of the verbal code. Paivio's theory is called a dual-coding theory because it proposes two independent memory codes, either of which can result in recall. A person who has stored both the word *butter* and an image of butter can remember to purchase the item if she retrieves either the image or the word. Evidence suggests that the two memory codes are independent in the sense that a person can forget one code without forgetting the other. Having two memory codes therefore provides a second chance of remembering an item.

The advantages of visual imagery for enhancing memory are evident from its emphasis in popular books for improving memory. Many authors of these books can perform impressive memory demonstrations such as repeating back the names of all members of the audience. The method used by the author Harry Lorayne (Lorayne & Lucas, 1974) involves first converting the name into a visual image and then linking the image to a prominent feature of the person's face. For example, if Mr. Gordon has a large nose, the image might be a garden (*garden* sounds like *Gordon*) growing out of his nose. Although the method may seem rather bizarre, it has experimental support.

Morris, Jones, and Hampson (1978) found that undergraduates at Lancaster University who were taught this strategy learned significantly more names than people who were not taught the strategy. The learning task required associating a different name (randomly selected from a telephone directory) with each of 13 photographs of male adults. After a study period of 10 seconds for each face, the imagery group could correctly name ten of the photographs, compared with five for the control group. The experimenters admitted that the use of visual strategies requires some effort, and not everyone will be willing to make that effort. However, demonstrations of the effectiveness of imagery for improving memory should provide encouragement to those who wonder whether the effort will be worthwhile.

ADVANTAGES OF IMAGES

Paivio's (1986) dual coding theory explains why two memory codes—visual and verbal—help us remember. However, another advantage of two memory codes is to reduce the interference between using memory and responding. Try to convince yourself of this benefit by performing the following pair of tasks. Look at the block diagram of the letter F in Figure 5.1 and form a visual image of the letter. Then mentally scan your visual image in a clockwise direction, beginning in the lower left corner. Say "yes" whenever you encounter a corner (right angle) that is on the top or bottom of the letter. Say "no" whenever you encounter a corner that is not on the top or bottom.

When you have finished that exercise, try this one. Remember the sentence "A bird in the hand is not in the bush." Now use your memory of the sentence to say "yes" whenever you encounter a word that is a noun and say "no" whenever you encounter a word that is not a noun.

For which task was responding more difficult? Why?

You may have experienced that it was more difficult to respond to words in the sentence because your verbal memory of the words interfered with your verbal response. In contrast, scanning a visual image allowed you to perform the task visually while you responded verbally.

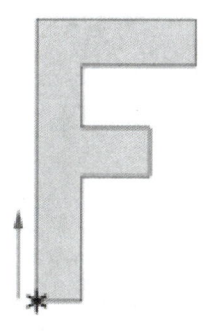

FIGURE 5.1 Block diagram of the letter F. (From "Spatial and Verbal Components of the Act of Recall," by L. R. Brooks, 1968, *Canadian Journal of Psychology, 22*, 349–368. Copyright 1968 by the Canadian Psychological Association. Reprinted with permission.)

What should happen if you had to respond visually by pointing to a *Y* for each positive response and to an *N* for each negative response? Research demonstrates that pointing interferes with the visual task. Classifying the words in a sentence took less time when people pointed to the correct response; classifying the corners of a letter took less time when people gave a verbal response. The advantage of having two memory codes—visual and verbal—is that designers can emphasize a memory code that does not interfere with responding.

There are, however, some unique advantages of a visual code. One difference between information maintained as a visual image and information maintained as a verbal code is that a visual image makes it possible to match information in parallel. When you look at the schematic faces in Figure 5.2, you can simultaneously perceive their features. However, when you describe these same features verbally you must describe them sequentially.

The parallel representation of visual information and the sequential representation of verbal information influence how quickly a person can determine whether a perceived pattern matches a memorized pattern. If the memorized pattern is stored as a visual image, the match should occur quickly and should be relatively uninfluenced by the number of features that have to be matched. If a pattern is stored as a verbal description, the match should occur more slowly and should be influenced by the number of features that have to be compared.

One of the first demonstrations of the advantages of parallel processing occurred in research by Nielsen and Smith (1973). Each trial in their experiment consisted of two sequentially presented stimuli separated by a short retention interval. The first stimulus was either a picture of a cartoonish face (Figure 5.2) or a verbal description of the face. The description represented the size of each feature such as "small nose," "large mouth," or "medium ears." Undergraduates at the University of Wisconsin had previously learned to associate these descriptions with pictures of the facial features. The second stimulus was always a picture.

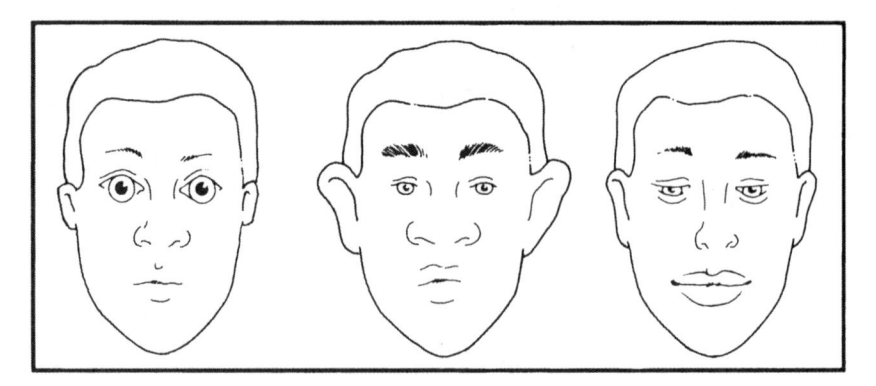

FIGURE 5.2 Cartoon faces used in a matching study. (From "Representation and Retrieval Processes in Short-Term Memory," by E. E. Smith and G. D. Nielsen, 1970, *Journal of Experimental Psychology, 10*, 438–464. Copyright 1970 by the American Psychological Association. Reprinted with permission.)

Producing Images 51

The task was to quickly respond "same" or "different" depending on whether this face matched the initially presented verbal description or picture. When both stimuli were pictures, subjects made very fast decisions and it did not matter whether they had to compare three, four, or five pairs of features. When the first stimulus was a verbal description, decisions were much slower and response times increased with the number of features that had to be compared. Although it would have been possible to describe verbally the features of the initial faces, the descriptions would have transformed the fast visual comparisons into slow sequential comparisons.

PROTOTYPICAL IMAGES

Evidence that we are capable of forming images to represent pictures of individual objects raises the question of how we represent *categories* of objects such as cats, bicycles, and chairs. You have seen many different examples belonging to each of these categories. It would be impossible to recall images of each cat, bicycle, or chair that you have seen.

Psychologists have typically studied categorization of patterns by having people learn a limited number of patterns belonging to two categories. Figure 5.3 shows an example from one of my own experiments (Reed, 1972) in which undergraduates at UCLA initially learned to classify the 10 faces into either Category 1 (top row) or Category 2 (bottom row). They then used their knowledge of the two categories to classify an additional 25 test faces that did not match any of the faces in the two categories. Figure 5.4 shows 3 of the 25 test faces classified by the students.

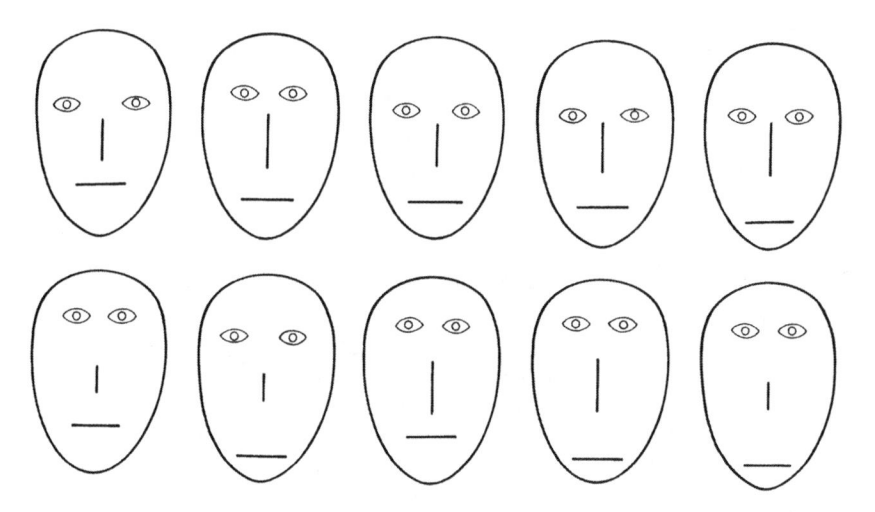

FIGURE 5.3 Two categories of schematic faces. From "Perceptual vs. Conceptual Categorization," by S. K. Reed and M. P. Friedman, 1973, *Memory & Cognition, 1*, 157–163. Copyright 1973 by the Psychonomic Society. Reprinted with permission.

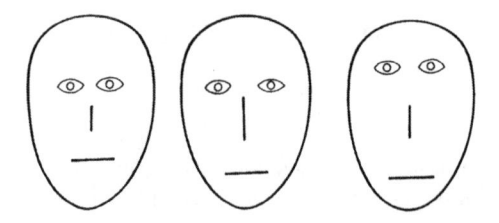

FIGURE 5.4 Test faces for categorization. (From "Pattern Recognition and Categorization," by S. K. Reed, 1972, *Cognitive Psychology, 3*, 382–407. Copyright 1972 by Elsevier. Reprinted with permission.)

The purpose of my research was to identify how people made their decisions. I used several measures including asking each person to select which of the four strategies listed in Table 5.1 best described their strategy (the strategies were not labeled as they are in the table). They most frequently selected the prototype strategy that involves creating an abstract image of a pattern to represent each category. I used the term "abstract image" to convey the idea that a prototype is *created* from the category patterns.

TABLE 5.1
Percentage of Subjects Who Reported Using Each Classification Strategy After Learning the Category Patterns in Figure 5.3

Strategy	Percentage
1. Prototype I formed an abstract image of what a face in Category 1 should look like and an abstract image of what a face in Category 2 should look like. I then compared the test face with the two abstract images and chose the category that gave the closest match.	58%
2. Feature frequency I looked at each feature on the test face and compared how many times it exactly matched a feature in each of the two categories. I then chose the category that gave the highest number of matches.	28%
3. Nearest neighbor I compared the test face with all the faces in the two categories, looking for a single face that best matched the test face. I then chose the category in which that face appeared.	10%
4. Average distance I compared the test face with each of the five faces in Category 1 and with each of the five faces in Category 2. I then chose the category in which the faces were more like the test face, basing my decision on all faces in the two categories.	4%

From: "Pattern Recognition and Categorization," by S. K. Reed, 1972, *Cognitive Psychology, 3*, 382–407. Copyright 1972 by Elsevier. Reprinted with permission.

I defined the prototype as the pattern that had the average feature values. The middle face in Figure 5.4 shows the Category 1 prototype consisting of the average eye width, eye separation, nose length, and mouth height of the five patterns in Category 1. The right face in Figure 5.4 shows the Category 2 prototype. A prototype would be represented spatially in the center of the category, as is illustrated by people's concept of a bird shown in Figure 4.5. A model that assumes people classify test patterns by comparing them to the two category prototypes made very good predictions of their classifications.

Although many people used a prototype strategy for this particular task, they appeared to use other strategies when the task changed. There was less evidence for the prototype strategy when people classified the test patterns by simply looking at the category patterns in Figure 5.3. When they did not have to learn the category patterns and rely on their memory, there was less motivation to simplify the task by creating prototypes. There was also less evidence for the use of a prototype strategy for a different pair of categories in which the two category prototypes looked very similar to each other.

Pani's (1996) functionalist position is helpful in providing an interpretation of these findings. He proposed that mental representations are produced according to the need for them and the need for imagery varies with the difficulty of the task. The need for creating a category prototype is greater when the examples have to be learned than when they are physically present. In addition, very similar prototypes increase the difficulty of using the prototype strategy and make it more likely that people will use an alternative strategy.

LIMITATIONS OF IMAGES

This chapter has surveyed several research studies that illustrate the advantages of visual imagery. Visual images can help us learn associations, reduce output interference, quickly match faces, and form prototypes to represent categories. After nearly half a century of neglect, research on visual imagery bounced back with new life fueled by the cognitive revolution.

However, I was concerned in the early 1970s that the pendulum had swung too far back in praise of the virtues of imagery. I had some nagging questions. Do images have limitations? Does language help support our use of images? Are images unable to be interpreted or do they have a structure? Does the Star of David contain a parallelogram (a four-sided figure with parallel opposite sides)?

The last question turned out to be a key for answering the first three questions. A standard embedded figures task requires finding a part (such as a parallelogram) in a whole (such as the Star of David) when both are simultaneously present. However, what if people had to examine an *image* of the whole to determine whether it contained a part? Could they correctly judge whether a visually displayed part was contained in a pattern they had just seen? This should be a trivial task if the part matched their verbal description of the whole. If you describe the Star of David as two overlapping triangles, then it will be easy to identify a triangle as a part. However, identifying a parallelogram as a part

FIGURE 5.5 An ambiguous figure. (From "Can Mental Images be Ambiguous?" by D. Chambers and D. Reisberg, 1985, *Journal of Experimental Psychology: Human Perception and Performance, 11*, 317–328. Copyright 1985 by the American Psychological Association. Reprinted with permission.)

requires examining a visual image. This turned out to be a difficult task. Only 10 of 80 undergraduates at Case Western Reserve University correctly identified that the second pattern was contained in the first when a parallelogram was shown immediately after the Star of David. They also had difficulty identifying parts in other images.

Here is an opportunity for you to beat the odds and demonstrate that your visual imagery is much better than average. Form a visual image of the animal in Figure 5.5. Now examine your image, without looking at the drawing in the book, and see if you can reinterpret the figure as a different animal. Were you successful? If not, try to reinterpret the figure in the book. If your experience was similar to the students studied by Chambers and Reisberg (1985), you should have found it much easier to reinterpret the drawing than to reinterpret your image. In fact, in one of their studies they found that none of the 15 students in the experiment could reinterpret their image but all 15 could reinterpret the drawing.

The *verbal overshadowing* effect provides one explanation of why it is difficult to reinterpret an image. Schooler and Engstler-Schooler (1990) discovered that producing verbal descriptions of visual stimuli could reduce the effectiveness of visual memories and impair subsequent recognition. People who were instructed to verbally describe pictures of faces were less able to subsequently recognize those faces than people who did not produce verbal descriptions. Brandimonte and Gerbino (1993) subsequently found that their undergraduates at the University of Trieste were more successful in reinterpreting an image of the duck/rabbit ambiguous figure if discouraged from forming a verbal description. Brandimonte and Gerbino concluded that people are better at maintaining the details needed for reinterpretation when relying solely on the image. As recommended by the Schoolers, some things are better left unsaid.

Fortunately, images that lack details can still be useful in many tasks such as remembering words. Your image of a rabbit would not have to be detailed or accurate to help you remember the word *rabbit;* it would only have to contain enough detail to allow you to recall the correct word. Studies showing the limitations of visual images provide both good news and bad news. The bad news is that using visual images is not a universal solution for improving memory performance. The good news is that even mediocre images may still be sufficient for performing the many tasks that do not require great detail. In addition, there is even better news for people with limited imagery. Vivid imagery can at times be more of a curse than a blessing when distinguishing between real and imagined events.

REALITY MONITORING

Like eating too many desserts, having very vivid images can be too much of a good thing. If our images of objects or events were as accurate and detailed as the actual events, then our ability to distinguish between actual and imagined events would be impaired. The ability to make this distinction has been called *reality monitoring* by Johnson and Raye (1981).

To study how well people can distinguish between actual and imagined events, Johnson, Raye, Wang, and Taylor (1979) showed students pictures and names of common objects. Undergraduates saw the picture of each object, two, five, or eight times and the name of each object, either two, five, or eight times. They were instructed to generate an image of the object each time its name occurred. At the end of the session, they received an unexpected test in which they had to estimate how often they had seen each of the pictures, ignoring the number of times they had generated images.

If people are very good at discriminating between seeing and imagining pictures, then their estimates of seeing should not be influenced by the number of times they imagined each picture. Note that although the ability to form accurate images is an asset for performing most spatial tasks, good imagery would be a liability for reality monitoring. Because people with good imagery might find it difficult to discriminate between what they imagined and what they saw, the experimenters gave participants an imagery test to measure their imagery ability.

The students were quite good at estimating how many times they had seen each picture, even when they had to distinguish between seeing and imagining. However, the image trials did influence their judgments—estimates of how frequently a picture occurred increased as the number of image trials increased. The number of image trials had a greater effect on the good imagers than on the poor imagers. As expected, people with good imagery distinguished less accurately between seeing and imagining.

Johnson and Raye (1981) proposed that several kinds of cues are helpful for distinguishing between perception and imagination. First, there is *sensory information*. Perceptual events have more sensory detail than imagined events, although the amount of sensory information is greater for good and than for poor imagers. Second, there is *contextual information*. Perceptual events occur in an external context that contains other information. For example, perceiving a clock is typically accompanied by the perception of other objects near the clock. Imagining a clock typically consists of an image of a clock that is not surrounded by other objects.

A third cue for distinguishing between perception and imagination is memory for the *cognitive operations* that are required to generate the image. When we generate an image automatically without much conscious awareness, we have poor memory of the cognitive operations used to generate the image. Dreams that occur during sleep are of this type. They often seem very real because we are not aware of generating them. In contrast, daydreams seem much less real because they are more influenced by our conscious control.

BREAKDOWN OF REALITY MONITORING

Johnson and Raye's (1981) use of the term "reality monitoring" has clinical implications because a breakdown in reality monitoring can result in hallucinations. Bentall's (1990) review of research on hallucinations indicates that they result from an impairment of skills in discrimination between real and imaginary events. The analysis of reality discrimination of normal people should therefore provide a valuable source of information for clinicians. However, according to Bentall, it is important to recognize that hallucinators do not imagine random events. The content of the hallucinations is presumably related to the personalities and the stresses of the patient. The challenge is to find which types of reality discrimination errors are linked to the different kinds of hallucinatory experiences.

A study in Wales (Sack, van deVen, Etschenberg, & Linden, 2005) found evidence that schizophrenics have enhanced imagery for all of the sensory modalities. The investigators gave a battery of tests to 50 schizophrenic patients and 50 age- and sex-matched healthy control participants. One of the tests was a standard questionnaire that measured the vividness of imagery on a 7-point scale ranging from "I perceive it perfectly clearly as if it were real" to "I think about it but I cannot imagine it." The questionnaire contained 35 examples (the smell of leather, the meowing of a cat) that included seven sensory modalities. The schizophrenics gave significantly higher vividness ratings for all sensory images in comparison to the control group.

Another very important aspect of reality monitoring concerns whether a traumatic event, such as childhood sexual abuse, actually occurred or was imagined, perhaps because a therapist or other authoritative person suggested that it occurred. The importance of this topic resulted in an entire issue of the journal *Applied Cognitive Psychology* devoted to it. The lead article by Lindsay and Read (1994) set the stage with these opening comments:

> There is no doubt that many children are sexually abused, and that this is a tragedy. Furthermore, survivors of childhood sexual abuse often suffer long-lasting harm, and may be helped by competent therapists. Although cognitive researchers have differing views about the mechanisms underlying loss of memory (e.g., repression, dissociation, or normal forgetting; see E. F. Loftus, 1993; Singer, 1990), all would agree that it is possible that some adult survivors of childhood abuse would not remember the abusive events, and that memories might be recovered given appropriate cues. Thus, we accept that some clients may recover accurate memories of childhood sexual abuse during careful, non-leading, non-suggestive therapies. But there is no doubt in our minds that extensive use of techniques such as hypnosis, dream interpretation, and guided imagery (which are advocated in some self-help

books and by some clinical psychologists, psychiatrists, clinical social workers, therapists, and counselors) can create compelling illusory memories of childhood sexual abuse among people who were not abused. This too is a tragedy. (pp. 281–282)

Lindsay and Read (1994) indicate that memory research has identified a number of factors that increase the possibility of creating false memories. These include long delays between the event and the attempt to remember, repeated suggestions that the event occurred, the perceived authority of the source of the suggestions, the perceived plausibility of the suggestions, mental rehearsal of the imagined event, and the use of hypnosis or guided imagery. Because some of these factors are necessarily part of therapy, practitioners need to be particularly concerned about the use of techniques that may increase the risk of creating illusory memories.

A study by Hyman and Pentland (1996) at Western Washington University demonstrated that the repeated use of guided imagery significantly increases illusory memory of childhood events. They asked parents of college students to provide a description of early events that their child might be able to recall. Students were then asked if they could remember the events and supply some additional information during a series of three interviews. In addition to the true events from their past, one of the events was false. Each participant was told, "You attended a wedding reception when you were 5 years old in which you spilled the punch bowl on the parents of the bride."

The experimenters compared an imagery condition with a control condition that differed in the advice given for aiding recall. When students in the imagery condition failed to recall information about either a true or a false event, they were instructed to imagine the event. When participants in the control condition failed to recall an event, they were required to think quietly about the event for 45 to 60 seconds. Events were categorized as recalled if the participant claimed to remember the event and provided additional information about what had occurred. By the end of the third interview, 12 of the 32 participants in the imagery condition falsely recalled the punch bowl incident, compared with 4 of 33 participants in the control condition.

Creating illusory memories can be especially damaging because there are no guaranteed techniques that experts can use to discriminate between real and false memories. Lindsay and Read (1994) argue that the overriding theme of their literature review is that illusory memories can look, feel, and sound like real memories and include the strong affect that accompanies real memories of childhood abuse. The experience of coming to believe that abusive events occurred would be tremendously traumatic regardless of whether the remembered event actually occurred.

SUMMARY

Producing visual images can be beneficial for many situations. Images can help us learn associations including names of people, reduce interference from language, simultaneously view the features of a pattern, and create category prototypes. However, images also have their limitations. Objects that are imagined are seldom as vivid, detailed, and accurate as objects that are perceived. This difference in the sensory quality of perception and imagination is necessary to enable us to distinguish between reality and thought. Fortunately, our imagery is usually good enough to allow us to improve our memory and thinking without being so good that it creates a breakdown in reality monitoring. The next chapter contains examples of how we can productively manipulate our images.

6 Manipulating Images

Look at each of the three pairs of patterns in Figure 6.1. For each pair, decide whether the two three-dimensional forms are the same object. Then reflect on how you made your decisions.

Shepard and Metzler (1971) designed this task at Stanford by creating some pairs (such as the top and middle ones) that were different orientations of the same pattern and other pairs (such as the bottom one) that were two different patterns. The pairs differed in orientation from 0° to 180°. Half of the pairs could be rotated to match each other, and half were mirror images that did not match.

Eight dedicated adults made same/different decisions for 1,600 trials. Their decision times increased linearly with an increase in the number of degrees the forms differed in orientation, suggesting that they were rotating a visual image of one of the forms until it had the same orientation as the other form. Self-reports were consistent with this interpretation—the participants reported that they imagined one object rotating until it had the same orientation as the other and that they could rotate an image only up to a certain speed without losing its structure.

This chapter discusses the mental manipulation of images through mental rotation, visual scanning, combining parts, solving problems, and designing products. These tasks add to our growing list of how images can help us.

SCANNING IMAGES

Stephen Kosslyn at Harvard has been the principle architect in building a theory of visual images. Kosslyn proposed in one of his early experiments that if perceptual operations on visual images are similar to perceptual operations on pictures, then people should scan images in the same way that they scan pictures (Kosslyn, Ball, & Reiser, 1978). He therefore predicted that the time to scan mentally between two objects in an image should depend on the distance between the objects. Kosslyn tested his hypothesis by requiring undergraduates at Johns Hopkins to learn the exact locations of the objects shown in Figure 6.2. He then removed the map and gave the students a series of trials that began with the name of an object. Participants then heard the name of a second object and followed instructions to scan their image of the island by imagining a black speck moving in a straight line from the first object to the second. When they reached the second object, they pushed a button that stopped a clock. Figure 6.3 shows that the average response times confirmed Kosslyn's prediction regarding the effect of the distance on mental scanning time.

These results suggest that we mentally scan visual images in the same way that we can scan pictures. However, critics of visual imagery such as Pylyshyn (1981) argued that participants in imagery tasks could respond appropriately

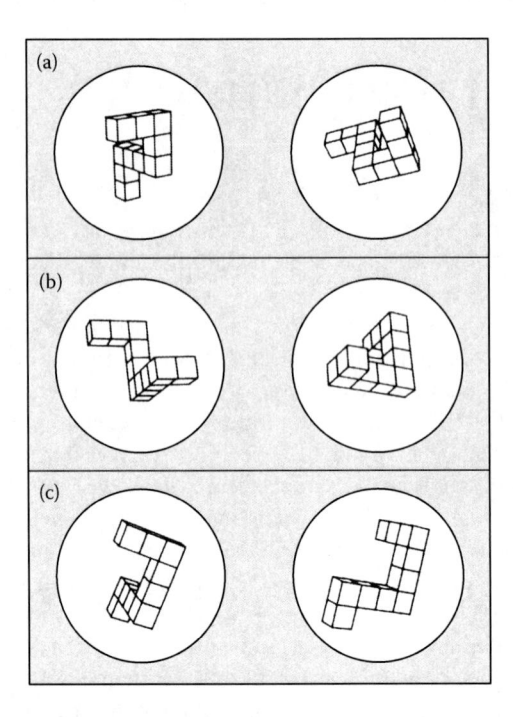

FIGURE 6.1 Pairs of complex patterns differing in orientation. (From "Mental Rotation of Three-Dimensional Objects," by R. N. Shepard and J. Metzler, 1971, *Science, 171*, 701–703. Copyright 1971 by the American Association for the Advancement of Science. Reprinted with permission.)

without actually using visual images. According to this view, participants knew that it would take longer to scan as the distances increased. It is therefore possible that they did not scan visual images but simply waited longer before pushing the response button as the distance increased between two objects. This argument gained support from results that showed people could accurately predict how distance should influence scanning time. When experimenters asked people to predict their scanning time for the different pairs of objects in Figure 6.2, predicted scanning times also increased as a linear function of distance (Mitchell & Richman, 1980).

The argument that people wait longer for longer distances did not seem plausible to me because it substitutes a harder task for an easier task. The easier task is simply to follow the instructions by mentally scanning from one object to another and pressing the response button when arriving at the second object. The harder task is to (1) estimate the distance between the two objects (such as 8 cm), (2) estimate the scanning time between the two objects (such as 1.3 seconds), and (3) press the response button when the estimated time has elapsed. Even this last step—determining when the estimated time has elapsed—seems more difficult than simply scanning the image.

Manipulating Images   61

FIGURE 6.2 A map of an island. (From "Visual Images Preserve Metric Spatial Information: Evidence from Studies of Visual Scanning," by S. M. Kosslyn, T. M. Ball, and B. J. Reiser, 1978, *Journal of Experimental Psychology: Human Perception and Performance, 4*, 47–60. Copyright 1978 by the American Psychological Association. Reprinted with permission.)

   My intuitions are not always correct, however, so I decided to do an experiment. The claim that people avoid mental scanning by predicting the outcome can be eliminated if the outcome of an experiment cannot be predicted. Reed, Hock, and Lockhead (1983) therefore attempted to design an experiment in which people could not predict the outcome. We believed that people might not be able to predict how the shape of a pattern would influence their scanning time. We therefore varied the length of three different shapes—a diagonal line, a curved version of a line shaped like an Archimedes spiral, and a maze-like version of the spiral consisting of lines joined at right angles (Figure 6.4). We expected that the fastest scanning times would occur for the diagonal line and the slowest scanning times would occur for the maze. Scan three patterns in the same column (which are the same length) to determine how the patterns influence your scan times.
   The scan times supported our predictions when people scanned the patterns and when they scanned visual images of the patterns. In both cases, the 24 patterns (3 shapes × 8 lengths) appeared on a screen in random order. In the perception condition, people scanned the pattern on the screen by pushing a button when they began their scan and pushed it again when they finished their scan. In the imagery condition, the pattern appeared on the screen for 0.5 seconds. People

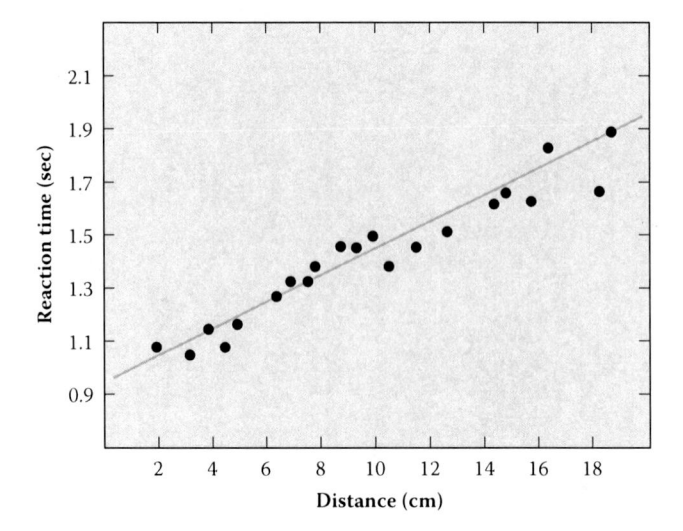

FIGURE 6.3 Time to scan between two objects on the island. (From "Visual Images Preserve Metric Spatial Information: Evidence from Studies of Visual Scanning," by S. M. Kosslyn, T. M. Ball, and B. J. Reiser, 1978, *Journal of Experimental Psychology: Human Perception and Performance, 4*, 47–60. Copyright 1978 by the American Psychological Association. Reprinted with permission.)

FIGURE 6.4 Three sets of patterns used to study mental scanning. (From "Tacit Knowledge and the Effect of Pattern Configuration on Mental Scanning," by S. K. Reed, H. Hock, and G. R. Lockhead, 1983, *Memory & Cognition, 3*, 569–575. Copyright by the Psychonomic Society. Reprinted with permission.)

then scanned a visual image of the pattern by pressing the button twice as they did in the perception condition. Figure 6.5 shows that the scanning times were remarkably similar for scanning both the patterns (Figure 6.5a) and images of those patterns (Figure 6.5b). The slope of the lines in Figure 6.5a and Figure 6.5b represents the rate of scanning, which was over twice as slow for the mazes as for the diagonal lines. Figure 6.5c shows that people were unable to predict how the three different patterns would influence their scanning times. The predicted times have essentially the same slopes. We therefore concluded from our results that

Manipulating Images

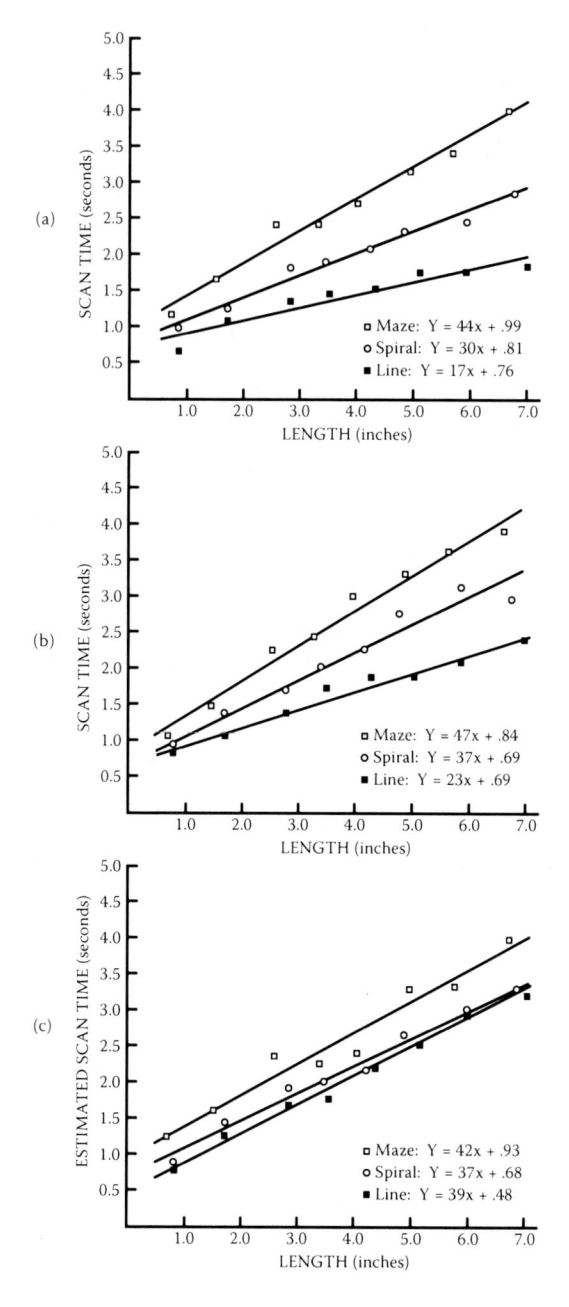

FIGURE 6.5 Scan times for perceived patterns (a), scan times for imaged patterns (b), and estimated scan times (c). (From "Tacit Knowledge and the Effect of Pattern Configuration on Mental Scanning," by S. K. Reed, H. Hock, and G. R. Lockhead, 1983, *Memory & Cognition, 3*, 569–575. Copyright by the Psychonomic Society. Reprinted with permission.)

people do scan visual images, and that scanning an image of a pattern is similar to scanning the actual pattern.

CAUSAL REASONING

A more challenging test of our imagery ability is making inferences by mentally simulating the operation of mechanical systems. The ability to mentally animate static pictures can be useful for solving spatial problems that require causal reasoning. Look at the diagram of the pulley system in Figure 6.6. Determine whether the pulley on the left turns in a clockwise or a counterclockwise direction when the rope on the right side of the diagram is pulled. You probably tried to mentally animate the pulley system to answer the question. The mental animation creates a causal chain of events that begins with moving the first (right) pulley, proceeds to moving the second (middle) pulley, and ends with moving the last (left) pulley. Hegarty (1992) at the University of California, Santa Barbara hypothesized that people mentally animate the pulleys in this stepwise manner in order to understand how the entire pulley system works.

To test this hypothesis, she asked students to quickly respond "true" or "false" to a statement that appeared to the left of the diagram. Figure 6.6 shows two examples to illustrate that some of the (static) statements did not mention movement, whereas other (kinematic) statements did mention movement. If people mentally animate the pulleys to evaluate the kinematic statements, their response times should increase as the length of the causal chain increases. This is what Hegarty (1992) found. Students most quickly verified statements about the first pulley and least quickly verified statements about the last pulley. Verification errors also increased from the beginning pulley to the end pulley. In contrast,

Static statement:
The upper left pulley is attached to the ceiling

Kinematic statement:
The upper left pulley turns counterclockwise

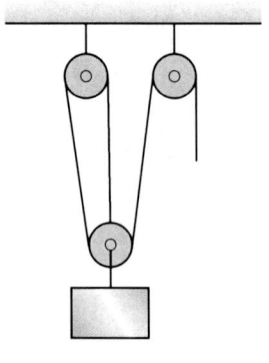

FIGURE 6.6 An example of a pulley problem. (From "Mental Animation: Inferring Motion From Static Displays of Mechanical Systems," by M. Hegarty, 1992, *Journal of Experimental Psychology: Learning, Memory and Cognition, 18*, 1084–1102. Copyright by the American Psychological Association. Reprinted with permission.)

Manipulating Images

neither response times nor errors varied across pulleys for the evaluation of the static statements. The location of the pulleys was only important when the verification task required their mental animation.

Schwartz and Black (1996) subsequently modified Hegarty's pulley task to determine if students' could use their mental simulations to discover rules for describing gear problems. An example problem is "Five gears are arranged in a horizontal line; if you try to turn the gear on the far left clockwise, what will the gear on the far right do?" Graduate students at Columbia University worked on a series of problems in which the number of gears in the chain varied. The experimenters used response times, error data, hand motions, and verbal reports of strategies to develop a model of the students' reasoning.

Figure 6.7 shows the model. Students initially simulate the task by using a detailed depiction of the gears. The fading phase creates a more schematic image by eliminating unnecessary pictorial details. The codifying stage uses language to describe the rule that gears alternate between moving clockwise and moving counterclockwise. The final quantitative reasoning stage relates the alternate rotation of the gears to a more general rule: The last gear in the chain will rotate in the same direction as the first gear if the number of gears is odd and will rotate in the opposite direction if the number of gears is even. Notice that discovering the rule eliminates the need for more laborious mental simulation. One simply has to count the gears to solve the problem.

The most complex mental simulation, to my knowledge, is the one described by Grandin (Grandin & Johnson, 2005) in her book *Animals in Translation*. Grandin has a doctorate in animal science, teaches at Colorado State University, and holds numerous patents. She is also autistic.

So how did Grandin become so successful? She attributes her success to the power of her ability to think in visual images. When Grandin thinks, she has no words in her head, only pictures. When she thinks about the structure of some design, she sees images of her design going together smoothly, images of problems and sticking points, or, if the design is flawed, images of the whole structure collapsing. Words occur to her only *after* she finishes her design.

When Grandin initially observed the operation of a meatpacking plant, she was amazed that anyone could understand and remember such a complex sequence of operations. She stood for hours on a catwalk overlooking the floor where approximately 100 employees processed the carcasses. Every Tuesday afternoon Grandin downloaded more visual details into her memory. After 24 Tuesday afternoons, she had a breakthrough:

> Then one day I was standing on the catwalk and suddenly it all seemed simple, I didn't have to worry about remembering the sequence anymore, because I could walk through the whole plant in my mind. Every step in the sequence was connected to the next step, so I didn't have to hold hundreds of different, separate details in my working memory at the same time. I just had to remember one step at a time, and that step brought up the next step…The breakthrough with the meat plant came when I could put the whole plant in one window and not have to switch

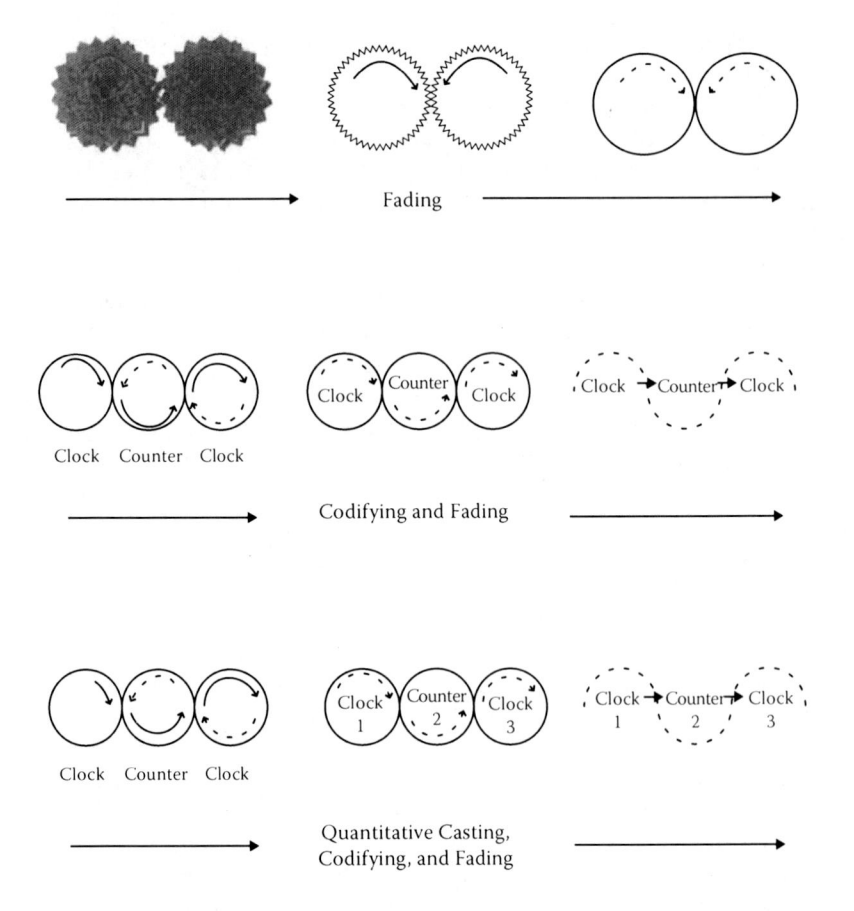

FIGURE 6.7 A model of transitions from simulations to rules. (From "Shuttling Between Depictive Models and Abstract Rules: Induction and Fallback," by D. L. Schwartz and J. B. Black, 1996, *Cognitive Science, 20*, 457–497. Copyright by Ablex. Reprinted with permission.)

> back and forth. Then I could understand and remember it, and after that when I visited other meat plants I could easily pick out the familiar machines even though the floor layout was different. (Grandin & Johnson, 2005, p. 253)

Grandin's achievement may be the most complex mental simulation ever created.

REARRANGING OBJECTS

The previous chapter briefly mentioned two theoretical movements—functionalism and behaviorism—that influenced the study of imagery. In 1910, the research of Max Wertheimer in Germany gave rise to another movement known as Gestalt psychology that would also have an impact. Perception was the core of Gestalt psychology and its central claim was that a pattern is more than

Manipulating Images

the sum of its parts. The view that images have structure fits nicely into this tradition.

Gestalt psychologists, including Wertheimer, subsequently applied their view of perception to problem solving. The problems required the rearrangement of objects in order to find the correct structural relation among the parts. A well-known example is the task described by Kohler (1925) in his book, *The Mentality of Apes*. Kohler hung some fruit from the top of a cage to investigate whether a chimpanzee or other ape could discover how to reach it. The cage also contained several sticks and crates. The solution depended on finding a correct way to rearrange the objects—for example, standing on a crate and using a stick to knock down the fruit. According to the Gestalt analysis, solving the problem required the reorganization of the objects into a new structure.

Gestalt psychologists argued that discovering the correct organization usually occurred as a flash of insight. *Insight* is the sudden discovery of the correct solution following a period of incorrect attempts based primarily on trial and error. The metaphor of the solution suddenly becoming visible is perhaps not surprising, based on the Gestalt interest in perception. The term "insight" itself emphasizes its parallel with vision, as do the expressions "a moment of illumination" and "seeing the light." The key factor distinguishing insight from other forms of discovery is the suddenness of the solution. In contrast to solutions that are achieved through careful planning or through a series of small steps, solutions based on insight seem to occur "in a flash."

An example of an insight problem studied by Metcalfe (1986) is the Gardener's Problem:

Gardener's Problem

A landscape gardener is given instruction to plant four special trees so that each one is exactly the same distance from each of the others. How should he arrange the trees?

Do not create a square because the diagonal of a square is longer than its sides. I include this problem in my cognitive psychology lectures because students typically cannot solve it without a hint. The hint for solving the problem is that you need to think in three dimensions.

One factor that can make it difficult to find a solution to an insight problem is that the problem solver places unnecessary constraints on the solution. According to this hypothesis, insight occurs when the problem solver removes the self-imposed constraint. The self-imposed constraint for the Gardener's Problem is that the trees must be planted on a flat surface. The solution requires planting the trees at the corners of a three-dimensional pyramid.

You can test whether you place unnecessary constraints on problems by trying to solve three Matchstick Problems.

Matchstick Problems

Move a single stick to turn a false arithmetic statement into a true statement. The stick cannot be discarded, but it can be rotated and it can change positions in the equation.

$$IV = III + III$$
(a)

$$IV = III - I$$
(b)

$$III = III + III$$
(c)

From: "Constraint Relaxation and Chunk Decomposition in Insight Problem Solving," by G. Knoblich, S. Ohlsson, H. Haider, and D. Rhenius, 1999, *Journal of Experimental Psychology: Learning, Memory and Cognition, 25*, 1534–1555. Copyright 1999 by the American Psychological Association. Reprinted with permission.

An experiment by Knoblich, Ohlsson, Haider, and Rhenius (1999) confirmed the hypothesis that matchstick problems of Type A would be the easiest and problems of Type C would be the most difficult. Type A problems are solved by moving a matchstick in the *Roman numerals*, such as turning the number IV into the number VI to make the statement VI = III + III. Problems of Type B are solved by changing the *arithmetic operations*, such as moving a stick from the equals sign to the minus sign to make the statement IV − III = I. Problems of Type C create two *equal signs*, such as III = III = III. Once people realize that they can modify and create new equal signs, the three types of problems become equally easy.

DESIGNING WITH IMAGES

It is unfortunate that cognitive psychologists have developed very few research programs to learn about the role of visual imagery in creative design, but much of what we do know was discovered by Finke, Ward, and Smith (1992) when they worked together at Texas A&M University. Finke had been one of the main

Manipulating Images

contributors to a theory of visual imagery, and he utilized his expertise in this area to extend imagery paradigms to study creative design.

His task required that people combine object parts to invent useful and novel products. The object parts consisted of such basic three-dimensional forms as a sphere, half sphere, cube, cone, cylinder, rectangular block, wire, tube, bracket, flat square, hook, wheels, ring, and handle. After either the experimenter or the participant selected three parts, the participants were instructed to close their eyes and imagine combining the parts to make a practical object or device. They had to use all three parts but could vary their size, position, and orientation. The created object had to belong to one of eight categories: (1) furniture, (2) personal items, (3) transportation, (4) scientific instruments, (5) appliances, (6) tools or utensils, (7) weapons, and (8) toys or games.

Image Design

Task	Parts	Category
1	Any three in Figure 6.8	Appliance
2	Half sphere, wheels, hook	Any of the eight
3	Cone, tube, ring	Game

You will have a better understanding of this task if you try the Image Design exercises. Select the appropriate parts from Figure 6.8, close your eyes, and combine the parts to make the designated practical object. Remember that you have to use all three parts but you can vary their size, position, and orientation.

Judges then scored the created objects on a 5-point scale for practicality and originality—the two essential components of creativity. In order for the completed object to be classified as creative, the average rating for practicality had to be at least 4.5 and the average rating for originality had to be at least 4.0. In one condition, participants were allowed to select their own parts but were told the category of their invention (such as appliances). In another condition, they were allowed to invent an object that belonged to any one of the eight categories but were told which parts to use (such as half sphere, wheels, and hook). In the most restrictive condition, they were told both the parts to use and the category of their invention (use a cone, tube, and ring to make a game). Although the number of inventions scored as practical was approximately the same across the three conditions, the most restrictive condition resulted in the highest number of creative inventions. The more restrictive the task for these three conditions, the more difficult it was to think of an object that was similar to existing objects.

There is, however, an even more restrictive condition than assigning both the parts and category. In another experiment, people were again given the parts, but did not find out about the category until *after* they had assembled their object. Finke (1990) referred to these objects as "preinventive forms" because

FIGURE 6.8 Object parts used to construct inventions. (From *Creative Imagery: Discoveries and Inventions in Visualization* (p. 41), by R. A. Finke, 1990, Mahwah, NJ: Lawrence Erlbaum Associates. Copyright 1990 by Lawrence Erlbaum Associates. Reprinted with permission.)

their use cannot be identified until after the object is constructed. Figure 6.9 shows how a preinventive form that was assembled from a half sphere, bracket, and hook could be used for each of the eight categories. The interpretations are lawn lounger (furniture), earrings (personal items), water weigher (scientific instruments), portable agitator (appliances), water sled (transportation), rotating

Manipulating Images 71

FIGURE 6.9 Multiple interpretations of a single preinventive form assembled from a halfsphere, bracket, and hook. (From *Creative Imagery: Discoveries and Inventions in Visualization* (p. 139), by R. A. Finke, 1990, Mahwah, NJ: Lawrence Erlbaum Associates. Copyright 1990 by Lawrence Erlbaum Associates. Reprinted with permission.)

masher (utensils), ring spinner (toys), and slasher basher (weapons). The judges' ratings indicated that this was the most successful condition of all for generating creative inventions.

THE GENEPLORE MODEL

Even in the less constraining conditions, many participants reported that they preferred initially to use a *generation strategy* in which they imagined interesting

```
          Geneplore Model

       GENERATE  ⟶  EXPLORE
               ⟵
            ↘    ↗
             CON-
           STRAINTS
```

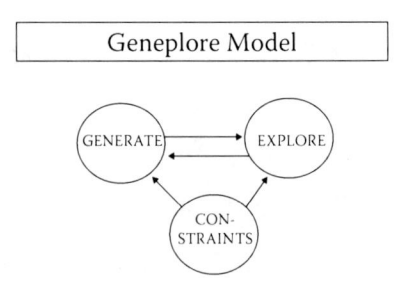

FIGURE 6.10 The structure of the Geneplore (generation–exploration) model. (From *Creative Cognition: Theory, Research, and Applications*, by R. A. Finke, T. B. Ward, and S. M. Smith, 1992, Cambridge, MA: The MIT Press. Reprinted with permission.)

combinations of parts, followed by an *exploration strategy*, in which they figured out how to use the invented object. Finke et al. (1992) describe these two phases in their Geneplore (generation–exploration) model shown in Figure 6.10.

In the initial phase, the inventor forms preinventive structures that are then explored and interpreted during the second phase. The preinventive structures are the precursors to the final, creative product and would be generated, regenerated, and modified throughout the cycle of invention. Finke et al. (1992) recommend that people should place greater emphasis on generating preinventive structures and then later think of possible uses. Notice that this is contrary to the usual order in which we begin with a particular use in mind—a wallet that fits in a shirt pocket or a garment that allows a swimmer to change clothes at the beach—and then try to invent something to accomplish our goal.

I was initially skeptical about the Geneplore model. It did not seem likely that designers assemble forms without having a particular goal in mind. Wouldn't they at least have a category in mind such as furniture, appliances, or transportation? I believe that designers typically begin with a general goal (such as design a car) but they can expand on that goal in creative ways by elaborating on their design structures. The Geneplore model could therefore apply to discoveries in this more constrained situation.

Hirshberg (1998) describes such a discovery in his book, *The Creative Priority*. As director of Nissan Design America, he studied the drawings of a car created by one of his designers. He noticed that an unusually thick line set off the upper rear quadrant of the car (including the rear window and trunk lid). The thick line separating this part of the car suggested to him that a vehicle could be decomposed into a set of interlocking panels that could be removed and replaced with other panels for diverse uses.

The designers went back to work to further develop this idea, while the engineers studied product constraints such as possible problems with hinges, structural reinforcements, and sealing. The various stages of the Geneplore model apply to the production of the Pulsar NX, the world's first production modular car. The preinventive structure in this case is a standard car, exploration of this structure leads to a modular car, and engineering constraints influence the final design.

Manipulating Images

SUMMARY

This chapter has shown how the manipulation of images applies to a variety of problem-solving tasks that involve mental rotation and causal reasoning. These tasks are mostly puzzles, but manipulating images is useful in daily activities such as imagining whether an object can fit in a car or whether a car can fit into a tight parking space. Manipulating images is also useful in many professions. Some of the items in the Dental Admissions Test, for example, require mentally rotating a complex shape to determine whether it could pass through a narrow opening. In addition, the manipulation of images is useful when combining parts to create products. You should be able to think of other instances in which manipulating images is beneficial in your daily or professional lives.

7 Viewing Pictures

Thinking visually often requires a visual display, such as a picture. Chapter 7 ("Viewing Pictures"), Chapter 8 ("Producing Diagrams"), and Chapter 9 ("Comprehending Graphs") discuss cognitive processes that we use to understand visual displays. The definitions of "picture," "diagram," and "graph" can vary, so let me clarify how I will use these terms. Pictures are illustrations that resemble the objects they depict, whereas graphs and diagrams are representations that are more abstract. Diagrams are abstract spatial layouts such as maps, Venn diagrams, and hierarchies. Figure 4.2 shows the distinction between a picture of containers and a Venn diagram of the containers. Graphs display data, such as changes in populations or fluctuations in stock prices.

Viewing pictures can affect our lives in many ways. At a personal level, pictures provide memories of important events such as birthday parties, weddings, and trips. At an emotional level, pictures can cause us to experience joy or sorrow, surprise or boredom, interest or disinterest. At a cognitive level, pictures can help us better understand ideas by providing illustrations of those ideas.

This chapter barely skims the surface of the many ways we experience pictures but it does provide some diverse examples. It begins with emotional responses to pictures by looking at people's reactions to political caricatures. It then discusses the possible tension between the affective and cognitive aspects of pictures by examining whether they support, or work against, each other. Although research studies have revealed that static pictures typically aid learning, animated pictures have produced mixed results. It is therefore important to determine when the extra time and money required to produce animation is justified—an issue raised at the end of this chapter.

EVOCATIVE PICTURES

Emotional pictures, whether pleasant or unpleasant, attract attention and do so very quickly, according to a study by Calvo at the University of La Laguna in Spain and Lang at the University of Florida (Calvo & Lang, 2004). They paired emotionally pleasant (affectionate) or unpleasant (threat or injury) pictures with unemotional control pictures and recorded eye movements to determine which pictures attracted viewers' attention. Both the probability of the initial fixation and the length of viewing times were higher for pleasant and unpleasant pictures than for neutral pictures during the first half-second of exposure. However, this initial attention to emotional pictures wore off rather quickly. Neither the probability of fixating nor the length of viewing time differed across pictures during the next 2.5 seconds of exposure.

I learned about the powerful evocative impact of pictures early in my research career. It occurred during a study of political caricatures with Mary Wheeler shortly after I joined the Psychology Department at Case Western Reserve in 1971 (Wheeler & Reed, 1975). Our interest in caricatures was influenced by our colleagues' interest in faces. As mentioned in Chapter 2, Fantz (1961) had been studying babies' attention to faces, Fagan (1973) was studying babies' memory for faces, and Cornell (1974) was studying babies' generalization of facial profiles. Wheeler and I were part of this movement. We had contributed our facial photographs to Cornell's project and I had recently published my 1972 article on categorization of schematic faces (Reed, 1972).

Wheeler initiated the question of whether political caricatures change over time to reflect the prevailing public opinion. The timing was right for studying this issue because President Richard Nixon's approval ratings were plummeting as a congressional committee headed by Senator Sam Ervin investigated the Watergate cover-up. The president's popularity was high in the latter months of 1972 as revealed by the results of a survey reported in *The Gallup Opinion Index*, Report No. 90 (December 1972). In response to the question, "Do you approve or disapprove of the way Nixon is handling his job as President?" 62% of the sample approved, 28% disapproved, and 10% expressed no opinion. In the first two months of the Watergate hearings (June and July 1973), the president's approval rating declined to 40%.

We began our study by collecting political cartoons that included the face of President Nixon. For each of eight leading political cartoonists (Conrad, Herblock, Interlandi, Luri, Mauldin, Oliphant, Osrin, and Szep), half of the cartoons were published in September or October 1972 before Watergate was a major issue and half were published in June or July 1973 when Watergate was a major issue. Figure 7.1 shows one of the later cartoons. We carefully cut out the faces of President Nixon and asked undergraduates to sort the caricatures for each cartoonist into two equal categories—a more favorable characterization of President Nixon and a less favorable characterization of President Nixon. The raters placed a significantly greater number of the 1973 post-Watergate caricatures in the less favorable category (Wheeler & Reed, 1975).

You may not think this finding is surprising but you have heard only half of the story. We still had many unused faceless cartoons and so we decided to repeat the experiment. The cartoons contained both the captions and the drawings, except for the missing faces. A new group of students sorted the cartoons of each cartoonist into a more favorable and less favorable characterization of President Nixon. There was no difference between the 1972 and the 1973 cartoons in judged favorableness. The declining approval rating of the president was reflected only in the facial caricatures, not in the caption and the context that surrounded the caricatures (Wheeler & Reed, 1975).

DECORATIVE PICTURES

If viewing pictures can have both a cognitive and an emotional effect, do these two effects work for or against each other? This, of course, is a complex issue and I want to examine only a small, but important, piece of it. You can probably think

FIGURE 7.1 Example of a political cartoon from the Wheeler and Reed (1975) study. (From "Response to Before and After Watergate Caricatures," by M. W. Wheeler and S. K. Reed, 1975, *Journalism Quarterly, 52*, 134–137. Copyright 1975 by Osrin. Reprinted with permission.)

of many instances in which you read material that was accompanied by interesting pictures. Interesting pictures may be a distraction if they do not contribute relevant information about the material. Levin (1989) has labeled such pictures "decorative illustrations" because they are entertaining but have no clear instructional value.

Harp and Mayer (1997) at the University of California, Santa Barbara investigated whether decorative illustrations would enhance learning a scientific text about lightning. The control group read a 550-word passage that included six black-and-white illustrations explaining the cause-and-effect steps in the formation of lightning. These illustrations were intended to have an instructional value. The decorative-illustrations group saw the same material as the control group with the addition of six color photographs taken from a *National Geographic* article on lightning. For instance, a photograph of swimmers accompanied the statement "Every year approximately 150 Americans are killed by lightning. Swimmers are sitting ducks for lightning because water is an excellent conductor of this electrical discharge."

Harp and Mayer (1997) found that students in the control group recalled more explanatory ideas than did students in the decorative-illustrations group. The authors concluded that their findings challenge the overuse in science textbooks of attention grabbing color photographs that are not directly relevant for making sense of an explanation.

Harp and Mayer's (1997) findings are important because many textbooks contain decorative illustrations that may be doing more harm than good. However, the findings also raise a broader question of when decorative illustrations should be included because Harp and Mayer refer to the interest generated by such illustrations. Students in psychology experiments are *required* to read text, but which variables influence their reading habits outside the laboratory? Many readers may be attracted to articles because of these attention-grabbing color photographs.

A particularly noteworthy photograph taken by Steve McCurry appeared on the June 1985 cover of *National Geographic*. The editor in chief, William Allen, reported that McCurry's photograph of an Afghan girl was the most memorable image that the magazine had ever published. During October 2005, the American Society of Magazine Editors reported a ranking of the top magazine covers of the last 40 years as judged by a panel of editors, artists, and designers. The *National Geographic* cover finished high in the competition.

Hidi and Renninger (2006) developed a model in which both affective and cognitive variables influence how people become interested in a topic. The four phases of their model are (1) triggered situational interest, (2) maintained situational interest, (3) emerging individual interest, and (4) well-developed individual interest. The two initial phases emphasize positive feelings for particular situations. The latter two phases emphasize the cognitive variables of values and knowledge to sustain this interest. A picture of a beautiful Afghan girl may trigger a person to read an article on the plight of Afghan refugees, but the person's values and knowledge are required to develop a sustained interest in this topic.

STATIC PICTURES

Decorative illustrations may be helpful for generating interest, but there is no doubt that informative pictures are valuable for helping readers understand text. Although Harp and Mayer (1997) found that decorative illustrations were distracting, they reported that illustrations explaining the cause-and-effect steps in the formation of lightning were beneficial. The best review that I have discovered on the benefits of visual representations in learning is a chapter by Anglin, Vaez, and Cunningham (2003) in the *Handbook of Research on Educational Communications and Technology*, second edition. The chapter includes a detailed summary of the instructional conditions, content of the passages, characteristics of the students, measures of learning, and the results of 90 studies on static pictures. The vast majority of the experiments found that text with explanatory illustrations was significantly more effective than text alone.

We should not assume, however, that novices view a picture in the same way that experts view a picture. A common obstacle for beginners is their failure to focus on the relevant details, as illustrated by the research of Lobato (1996, 2008). She studied a ninth-grade curriculum that spent considerable time on teaching the concept of slope. The instruction included a week of activities in which staircases were used to calculate slope as a vertical change (*rise*) divided by a horizontal change (*run*). Students also were taught how to find the slope of a line on a graph. They did well in calculating slope at the end of the instruction, scoring 80% correct on lines and 87% correct on staircases. However, only 40% of their answers were correct when they had to calculate the slope of a familiar object (a playground slide) that was not presented during the instruction.

Figure 7.2 shows that many incorrect responses were caused by students' difficulty in identifying which part of the picture should be the rise and which part should be the run. Some of these errors resulted from their inability to transfer relevant information from one object to another. For instance, those students who thought the length of the horizontal platform on the slide provides the *run* measurement incorrectly generalized from the horizontal lengths of the steps in the staircase.

The challenge of reasoning from previously presented pictures becomes clear from work on using artificial intelligence to design learning aids. Kenneth Forbus and his Qualitative Reasoning Group at Northwestern University are developing *Companion Cognitive Systems* to help people solve challenging problems that require reasoning from pictures (Forbus & Hinrichs, 2006). Forbus' group is evaluating the success of their software in solving problems on the Bennett Mechanical Comprehension Test, which has been administered for over 50 years to evaluate job candidates. Figure 7.3 (top) shows a typical problem that requires determining which crane is more stable. The *Companion* attempts to answer this question by searching its store of example problems to find a helpful solution.

Finding helpful examples requires the computer to search its memory for other drawings of cranes, but these drawings will not be identical to the ones in Figure 7.3 (top). As was the case for Lobato's (2008) students, it needs to transfer relevant pictorial details from one drawing to another. The *Companion* must

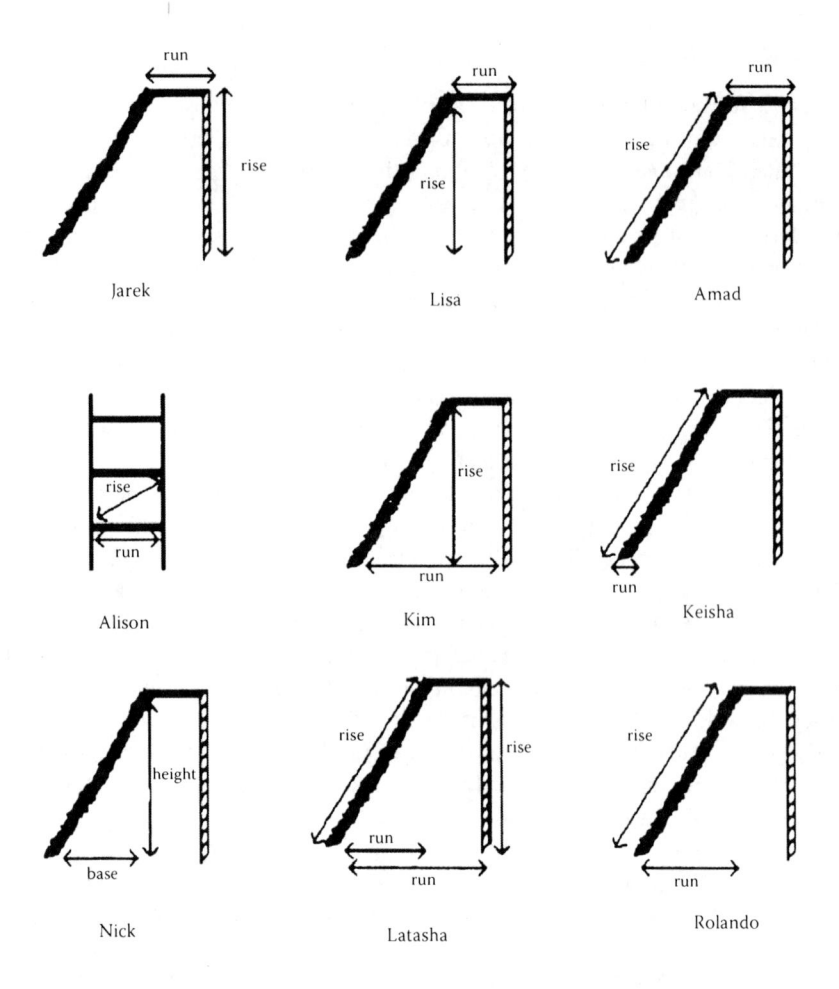

FIGURE 7.2 Finding the slope of a slide. (From *Transfer Reconceived: How "Sameness" Is Produced in Mathematical Activity*, by J. Lobato, 1996, unpublished doctoral dissertation. University of California, Berkeley. Reprinted with permission.)

also transfer relevant causal information across drawings to solve problems. To accomplish this task, its knowledge base includes over 38,000 concepts, over 8,000 relations, and over 5,000 logical functions. It does not always succeed but it was successful in constructing the causal model shown in Figure 7.3 (bottom). The causal reasoning from stored examples determines that it is necessary to focus on the distance from the cab to the load to find stability. The software therefore constructs lines to measure this distance and decides the crane on the right is more stable.

Carefully comparing pictures can also be important when students draw pictures. Van Meter's (2001) research at Penn State University investigated how providing support to fifth- and sixth-grade students helped them use their drawings to

Viewing Pictures 81

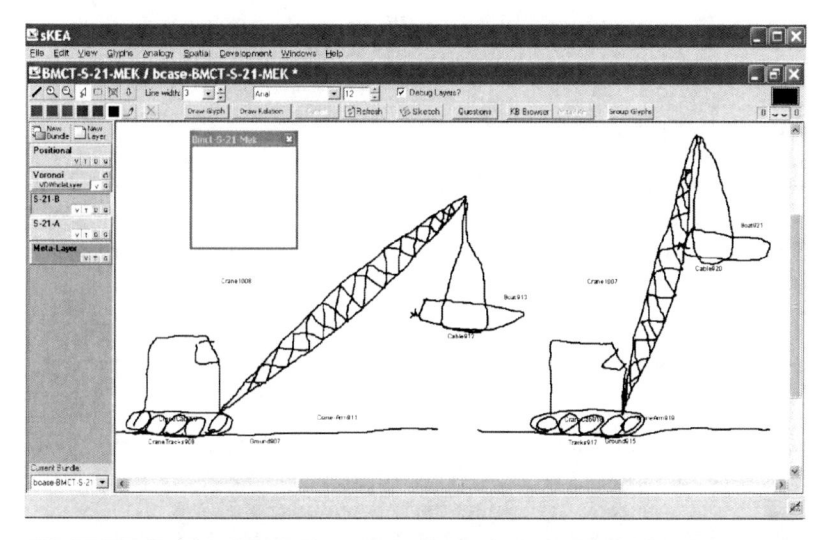

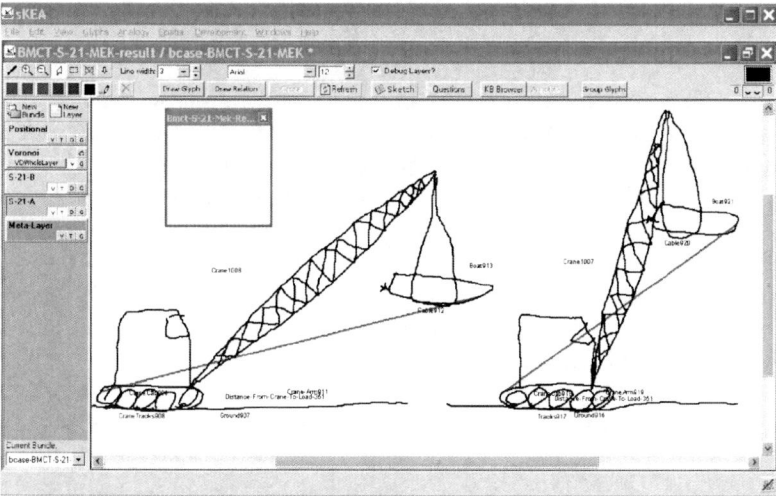

FIGURE 7.3 (Top) An example problem and (bottom) solution constructed by the *Companion Cognitive Systems* software. (From "Companion Cognitive Systems: A step towards human-level AI." by K. D. Forbur and T. R. Hinrichs, 2006, *AI Magazine*, 27, 83–95. Copyright by the American Association for Artificial Intelligence. Reprinted by permission.)

organize facts about the central nervous system. The groups receiving the most support made drawings as they read the text and then were prompted on how to modify their drawings while comparing them to the correct ones. Notice that viewing pictures is an important component of this procedure because of the requirement to compare drawings. Two other experimental groups were not prompted to systematically compare their drawings to the correct ones. Only the group who compared its drawings with the correct illustrations recalled more ideas than a control group who read the text with the correct illustrations but did not make drawings.

ANIMATED PICTURES

One method for directing attention to the relevant parts of a picture is to move those parts. A study by Grant and Spivey (2003) initially recorded eye movements to determine which parts were most relevant for solving a classical problem formulated by Gestalt psychologist Karl Duncker.

Duncker's Radiation Problem

Given a human being with an inoperable stomach tumor and lasers that destroy organic tissue at sufficient intensity, how can one cure the person with these lasers and, at the same time, avoid harming the healthy tissue that surrounds the tumor?

Figure 7.4 shows the picture that accompanied the problem, except the features were described verbally rather than labeled. The solution requires the use of multiple laser beams that converge on the tumor. The intensity of each individual beam is insufficient to harm healthy tissue but the combined intensity at the point of convergence is sufficient to destroy the tumor. Grant and Spivey (2003) hypothesized that unsuccessful solvers would focus more on the tumor and successful solvers would focus more on the skin because the multiple lasers must be imagined at the outer regions. The eye movements of their Cornell University undergraduates confirmed their hypothesis.

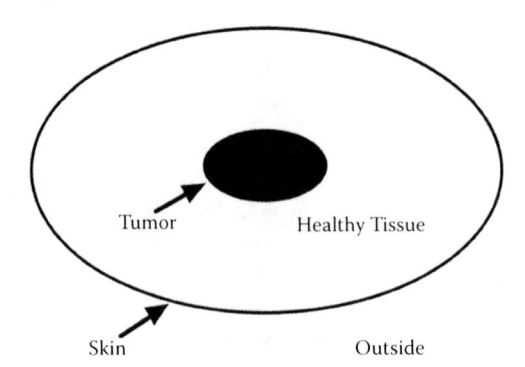

FIGURE 7.4 The picture accompanying Duncker's radiation problem in the Grant and Spivey (2003) study. (From "Eye Movements and Problem Solving: Guiding Attention Guides Thought," by E. R. Grant and M. J. Spivey, 2003, *Psychological Science, 14*, 462–466. Copyright by Blackwell. Reprinted with permission.)

A second, follow-up experiment capitalized on this finding by modifying Figure 7.4 to facilitate the correct solution. Students in the animated-skin condition viewed a picture in which the thickness of the skin subtly pulsed to attract attention. In contrast, students in the animated-tumor condition saw the tumor pulse. Students in the static condition viewed the static picture from the first experiment. Two-thirds of the students in the animated-skin condition solved the problem, compared to approximately one-third of the students in the animated-tumor and static conditions. These results are interesting because the importance of shifting attention to the skin is not intuitively obvious. Studying the eye movements of successful problem solvers was clearly helpful.

Although Grant and Spivey (2003) successfully used animation to attract attention, they may have been as successful in using color or some other cue in a static picture. Their findings therefore raise the issue of identifying when animated pictures are more helpful than static pictures. Let us return to the literature review in search of an answer.

Anglin et al. (2003) also reviewed studies that used visual animation to improve learning. These experiments used a variety of visual displays such as animated illustrations, motion, and interactive maps. The content areas included general science, physics, geometry, mathematics, statistics, and electronics. The learners ranged from primary-school students to adults. The tests measured (1) learning of facts, concepts, and procedures, (2) problem solving and visual thinking, and (3) acquisition of spatial skills. The outcome of these studies was more mixed than the positive gains typically found for static images. Significant animation effects were found in 21 of 45 comparisons.

The reviewers reported several specific recommendations based on the mixed results. The first is that more attention needs to be directed at the functional role of animation. They cite the conclusions from an earlier review by Park and Hopkins (1993) that state the successful designs of animation have identified those situations in which animated instruction is likely to be effective, such as depicting the motion or trajectory of an object. A recommendation made by Rieber (1990) following his 1990 literature review is that the greatest potential of animation comes from interactivity in which participants can control the animation, rather than simply observe. However, to assure that the animation is producing an instructional effect, Anglin et al. (2003) propose that it is necessary to show that the effect would not have occurred without motion.

Animated pictures are not always necessary because Hegarty's (1992) research on pulley problems and Schwartz and Black's (1996) research on gear problems showed that people could sometimes make causal inferences from *mental* simulations of static pictures. More recently, Hegarty, Kriz, and Cate (2003) conducted several experiments to determine if undergraduates could make more complex causal inferences about the operations of a flushing cistern (Figure 7.5). Although students' understanding was improved by viewing both static and animated diagrams, there was no evidence that the animated diagrams led to better understanding. The researchers proposed that the students were able to use their powers of spatial visualization and mental animation to understand the information provided in the text and the static diagram. They concluded that it is important to understand the powers and limitations of people's internal visualizations in order to best use our increasing power to generate animated instruction.

STATIC VERSUS ANIMATED PICTURES

Hegarty, Kriz, and Cate's (2003) conclusion is important because it would be a waste of time and money to build instructional animations of devices and situations that people can mentally simulate. Our Animation Tutor group at San Diego State is therefore conducting research to determine when instructional animations help students with their mathematical reasoning. One type of help is to provide visual feedback of students' estimates so they can improve their estimates.

A study conducted by Reed and Hoffman (2004) illustrates our experimental approach. Figure 7.6 shows four variations of a problem in which students estimate how long it would take two pipes to fill a tank. For example, an estimate of 3.6 hours for Problem A would fill approximately 60% of the tank and provide information that a student could use to revise her estimate. An effective strategy in this case is to use proportional reasoning: If 3.6 hours fill .6 tanks, how many hours will it take to fill 1.0 tanks? We gave both static and animated feedback to our San Diego State undergraduates to determine if animated feedback produced estimates that are more accurate. The static pictures were identical to the ones

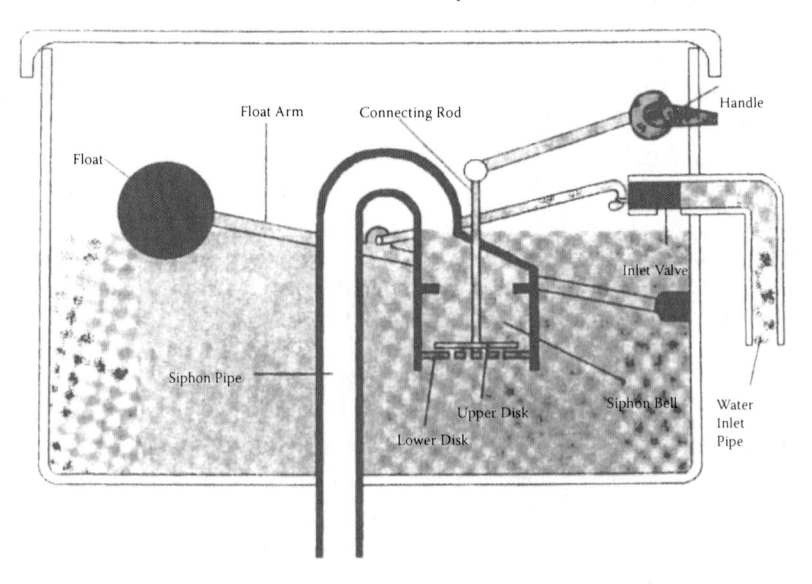

FIGURE 7.5 Static picture of a flushing cistern. (From "The Roles of Mental Animations and External Animations in Understanding Mechanical Systems," by M. Hegarty, S. Kriz, and C. Cate, 2003, *Cognition and Instruction, 21*, 325–360. Copyright by Lawrence Erlbaum Associates. Reprinted with permission.)

A

One pipe can fill a 10-foot tank in 10 hours and another pipe can fill a tank in 15 hours. How long will it take to fill the tank?

Time: 3.6 Hours

B

One pipe can fill a 10-foot tank in 10 hours and another pipe can fill a tank in 15 hours. There is a leak in the bottom of the tank that empties 1/18th of the tank each hour. How long will it take to fill the tank?

Time: 5.4 Hours

C

One pipe can fill a 10-foot tank in 10 hours and another pipe can fill a tank in 15 hours. There is a leak 3 feet from the bottom of the tank that empties 1/18th of the tank each hour. How long will it take to fill the tank?

Time: 4.5 Hours

D

One pipe can fill a 10-foot tank in 10 hours and another pipe can fill a tank in 15 hours. How long will it take to fill the tank if the 15-hour pipe begins 3 hours after the 10-hour pipe?

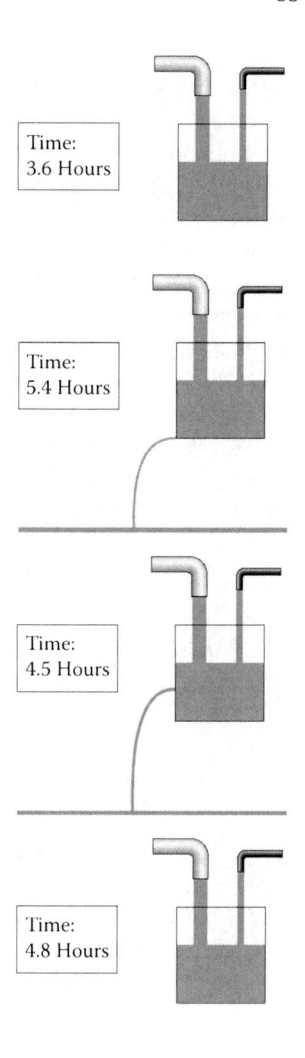

Time: 4.8 Hours

FIGURE 7.6 Tank problems investigated by Reed and Hoffman (2004). (From "Use of Temporal and Spatial Information in Estimating Event Completion Time," by S. K. Reed and B. Hoffman, 2004, *Memory & Cognition, 32*, 271–282. Copyright by the Psychonomic Society. Reprinted with permission.)

in Figure 7.6 and the animated pictures showed the tanks filling to the levels in Figure 7.6.

We hypothesized that the static pictures would be as effective as the animated pictures if there were no change in the rate of fill. The proportional reasoning strategy based on a static picture is a perfectly good strategy for Problems A and B in which the fill rate does not change. However, it requires adjustments when the fill rate changes as it does in Problems C and D. Rate of fill slows down in the side-leak problems (C) when the liquid rises above the leak and speeds up in the delay problems (D) when the second pipe begins to fill. We therefore predicted that the animated displays would be more effective than static displays when there were changes in the fill rate. Students' ratings supported our predictions that they focused more on the animation when there were changes in the rate of fill.

Another aspect of our research that is particularly relevant in this context is the accuracy of students' mental simulations. We have seen that students' mental simulations have supported causal reasoning, but could their simulations support accurate estimates of time? One of our tasks required that students mentally simulate the continued filling of the tank as soon as the animation stopped. They were told to use the animation time shown on the screen (as in Figure 7.6) and the additional time from their mental simulation to estimate the total time to fill the tank. We measured the accuracy of their mental simulations by requiring them to press a response key when their simulations reached the top of the tank. Their estimated completion times were much more accurate than their mental simulation times, indicating that they were using strategies other than mental simulation to improve their estimates. We were pleased that we had not wasted time and money on building instructional animations that could be mentally simulated.

SUMMARY

There is much evidence that explanatory static pictures aid learning. Decorative pictures can distract attention from explanatory information but may trigger situational interest. The effective use of static pictures requires directing attention to the most relevant parts of a picture. This may require comparing pictures, as when learning from previous examples or evaluating whether constructed pictures are correct. Animation can direct attention but the effectiveness of animation varies greatly across tasks. The challenge for researchers is to identify when instructional animations are required because mental simulations are inadequate.

8 Producing Diagrams

I distinguished between pictures and diagrams in Chapter 7 by describing diagrams as representations that are more abstract than pictures. I then showed you an illustration in Figure 7.4 for Duncker's radiation problem and called it a picture. The boundaries between linguistic categories are often fuzzy and you may prefer to call it a diagram. More important than the definitions, however, is recognition that both pictures and diagrams vary in detail and the amount of required detail depends on our goal.

Figure 7.4 provides sufficient detail for *discovering* the convergence solution to the radiation problem, but more detail is needed to *implement* it. Fortunately, technological advances can now provide that detail. In October 1998, UCLA's Jonsson Cancer Center announced a new treatment that enables doctors to create beams of radiation that exactly fit a tumor's dimensions. The system focuses thin rays of radiation through healthy tissue that converge on the tumor like spokes meeting at the center of a wheel. Solving puzzles occasionally results in marvelous inventions.

The need for detail also varies in our daily lives. Sometimes we need only a freeway map; at other times, we need a detailed street map. One of the assignments in Mark Harrower's Graphic Design in Cartography class at the University of Wisconsin requires taking a digital map of New Orleans and reducing it to fit into a 5-inch square. Students struggle with fitting the details, such as bends in a river and long road names, into the smaller space. They have to decide what elements to keep and how to illustrate them so that a viewer can understand the map.

Producing too much detail in larger maps can also create problems as was cleverly pointed out by Lewis Carroll, the author of *Alice's Adventures in Wonderland*. In Carroll's story, *Sylvie and Bruno Concluded,* the narrator meets a person by the name of Mein Herr, with whom he has the following conversation:

> "That's another thing we've learned from *your* Nation,'" said Mein Herr, "map making. But we've carried it much further than you. What do you consider the *largest* map that would be really useful?"
> "About six inches to the mile."
> "Only six inches!" exclaimed Mein Herr. "We very soon got to six yards to the mile. Then we tried a hundred yards to the mile. And them came the grandest idea of all! We actually made a map of the country, on the scale of a *mile* to the *mile*!"
> "Have you used it much?" I enquired.
> "It has never been spread out, yet," said Mein Herr, "the farmers objected: they said it would cover the whole country, and shut out the Sunlight!" (Norretranders, 1998, p. 399)

CONSTRUCTING REPRESENTATIONS

The organization of this chapter follows a developmental sequence from children to undergraduates to experts. Andy diSessa at the University of California, Berkeley is an expert who studies children. His current project is called Project MaRC, an acronym for metarepresentational competence. diSessa's research group is interested in how students

1. invent or design new representations,
2. critique and compare the adequacy of representations and judge their suitability for various tasks,
3. understand the purposes of representations and how they work for us,
4. explain representations, and
5. learn new representations quickly and with minimal instruction.

The MaRC group has focused on the construction of representations in science, particularly diagrams. One challenge for constructing a good representation is to determine how the use of space in a diagram can represent space in the world. The diagram in Figure 8.1 was constructed by a sixth-grade student to show the speed of a car by the length of arrows. The car slows to a stop (single dot), backs up (reverse arrows), waits a long time (three dots), and then accelerates away. Notice that the only problem with the diagram is the difficulty in using a single number line to represent backing up. The car appears to jump ahead before it backs up. The problem occurs because the single line must represent both space and time. Backing up could be represented by drawing backward arrows over the forward arrows, and then more forward arrows to again show moving forward. However, the diagram would be messy and hard to interpret.

Part of the ability needed to *produce* good representations is the acquisition of criteria for *judging* good representations. diSessa (2004) has found that even sixth-grade children have a rich set of criteria that include

- Completeness (shows all relevant information)
- Compactness (better use of space)
- Precision (you can more accurately read out quantitative information)
- Systematicity (obeys simple rules of correspondence)
- Conventionality (does not violate accepted conventions)
- Learnability (easy to explain)

FIGURE 8.1 A sixth-grade student's diagram of motion. From "Metarepresentation: Native Competence and Targets for Instruction," by A. A. diSessa, 2004, *Cognition and Instruction*, 22, 293–331. Copyright 2004 by Lawrence Erlbaum Associates. Reprinted with permission.

Producing Diagrams   89

Students, of course, gradually become more sophisticated in applying these criteria to construct good diagrams. One of the major developmental trends discovered in diSessa's studies is the movement away from realism to abstractness. This is particularly necessary for constructing scientific diagrams. Students gradually become more adept at using representations to highlight spatial relationships rather than physical details.

EMPHASIZING SPATIAL RELATIONS

The importance of spatial relationships is illustrated by Diezmann and English (2001) at the Queensland University of Technology. Figure 8.2 shows the effective use of a diagram by 10-year-old Kate. Kate represents the rows of bricks as a vertical number line and the movement of the frog by a series of arrows. Kate's diagram is more successful in representing both space and time than the diagram in Figure 8.1. She represents time as a second (horizontal) dimension that shows successive attempts to climb the wall. Kate's diagram provides a clear representation of the problem's structure and results in a successful solution.

In contrast, Kate's classmate Helen is unable to create an effective diagram for solving a similar problem. Helen's diagram, shown in Figure 8.3, does not include a number line and fails to show the movement of the koala. She has focused too much on including the pictorial details of the problem.

This distinction between schematic representations that show spatial relations and pictorial representations that show visual appearances was also studied by Hegarty and Kozhevnilov (1999). They gave spatial problems, such as the Path Problem, to 33 boys in a sixth-grade class in Dublin, Ireland.

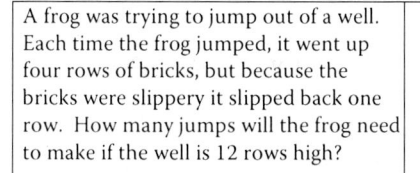

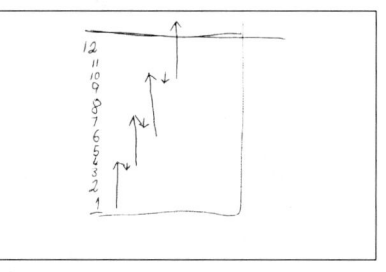

A frog was trying to jump out of a well. Each time the frog jumped, it went up four rows of bricks, but because the bricks were slippery it slipped back one row. How many jumps will the frog need to make if the well is 12 rows high?	

FIGURE 8.2 Kate's diagram of the frog problem. From "Promoting the Use of Diagrams as Tools for Thinking," by C. Diezmann and L. D. English, 2001. In A. A. Cuoco and F. R. Curcio (Eds.), *The Roles of Representation in School Mathematics* (pp. 77–89), Reston, VA: National Council of Teachers of Mathematics. Copyright 2001 by National Council of Teachers of Mathematics. Reprinted with permission.

| A sleepy koala wants to climb to the top of a gum tree that is 10 meters high. Each day the koala climbs up 5 meters, but each night, while asleep, slides back 4 meters. At this rate how many days will it take the koala to reach the top? | *[diagram of tree, 10m]* | Helen: "He climbs up 5 meters to there (5 meter mark) and that took him one day and that took him back down to here (just below the 5 meter mark) and he had to climb up another 5 (meters) the next day and he got about here ..." |

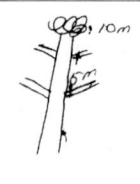

FIGURE 8.3 Helen's diagram of the koala problem. (From "Promoting the Use of Diagrams as Tools for Thinking," by C. Diezmann and L. D. English, 2001. In A. A. Cuoco and F. R. Curcio (Eds.), *The Roles of Representation in School Mathematics* (pp. 77–89), Reston, VA: National Council of Teachers of Mathematics. Copyright 2001 by National Council of Teachers of Mathematics. Reprinted with permission.)

Path Problem

At each end of the two ends of a straight path, a man planted a tree, and then every 5 meters along the path he planted another tree. The length of the path is 15 meters. How many trees were planted?

The researchers scored the representations of the problems as either primarily schematic or as primarily pictorial. Schematic representations focused on the spatial relations between objects, such as the distances between the trees, whereas pictorial representations focused on the objects themselves. Use of schematic representations had a positive effect on solving problems and use of pictorial representations had a negative effect. Hegarty and Kozhevnilov (1999) proposed that instructing students to "visualize" mathematical problems would not be successful unless it is clear that the visual representations should not contain irrelevant pictorial details.

Diezmann and English (2001) wrote their chapter on promoting the use of diagrams to help teachers help their students. Among their recommendations are the following:

- Ensure that the tasks are sufficiently challenging to warrant the use of diagrams.
- Actively promote the use of diagrams by modeling and discussing their use.
- Emphasize similarities and differences between problem structures.
- Encourage students to generate diagrams as a means for improving their understanding of problem structure.

REPRESENTING SPACE AND TIME

The amount of encouragement that people need to use a diagram depends on the task. Research indicates that we need very little encouragement if the task is to design a floor plan. However, much more encouragement is required if the project is to design the order of stages in a manufacturing process.

Carroll, Thomas, and Malhotra (1980), then working at the IBM Watson Research Center, studied the role of diagrams in design by creating two problems that had a similar list of constraints. The *spatial version* involved designing a business office for seven employees. Each employee was to be assigned to a corridor a certain number of offices down from a central hallway containing a reception area at one end and accounting records at the other end. College students were told to try to assign (1) employees who use accounting records closer to the accounting records, (2) compatible employees to the same corridor, and (3) employees with higher prestige nearer to the central hallway. The problem was accompanied by constraints describing these relationships between 19 pairs of employees. A general goal of the problem was to minimize the number of corridors.

Problems of this kind are usually easier to solve by using a diagram such as shown in Figure 8.4. The top of the diagram shows the central hallway connecting the reception and accounting areas. The columns represent corridors. Examples of constraints that are satisfied by the arrangement in Figure 8.4 are that A uses the accounting records less than C, B and C are compatible, and C has more prestige than B.

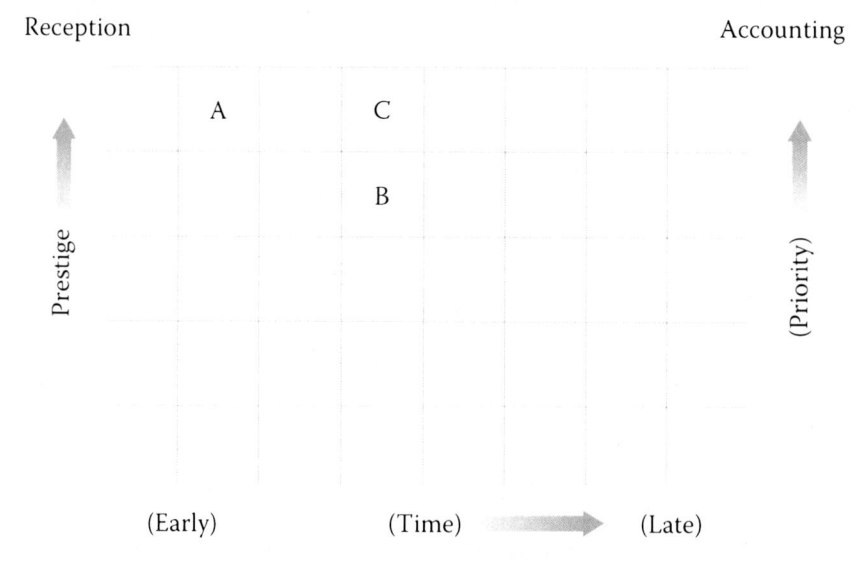

FIGURE 8.4 A matrix for representing a design problem. (From "Presentation and Representation in Design Problem Solving," by J. M. Carroll, J. C. Thomas, and A. Malhotra, 1980, *British Journal of Psychology, 71*, 143–153. Copyright 1980 by The British Psychological Association. Reprinted with permission.)

The *temporal version* of the problem had 19 equivalent constraints, but the constraints were placed on a manufacturing process that consisted of seven stages. The columns in Figure 8.4 can now be used to represent work shifts rather than corridors. The horizontal dimension represents time, and the vertical dimension represents priority. This group of designers was instructed to assign stages to the same work shift if the stages used the same resources. Some stages had to be assigned to earlier work shifts than others, and some stages had priority over others that belonged to the same work shift. Examples of constraints that are satisfied by the arrangement in Figure 8.4 are that stage A occurs before stage C, stages B and C use the same resources, and stage C has greater priority than stage B. Notice that distance from the accounting area in the spatial version corresponds to time in the temporal version, compatibility corresponds to use of the same resources, and prestige corresponds to priority.

The designers were not given a diagram or instructed to use a diagram; they were able to select their own method for solving the problem. Performance was measured by how many constraints were satisfied in the design. The importance of representation is illustrated by the finding that participants did significantly better on the spatial version even though the two versions had equivalent constraints. Designers given the spatial version not only satisfied more of the constraints but completed their design faster. All 17 participants in the spatial task used a sketch of the business office to formulate their design, but only 2 of the 18 participants in the temporal task used a diagram.

To determine whether a diagram made the problem easier for the spatial group, the experimenters conducted a second experiment, in which both groups were instructed to use the matrix shown in Figure 8.4. This time there were no significant differences between the two groups, either in performance scores or in solution times. The differences in the first experiment therefore appear to have been caused by the facilitating effects of a diagram. The usefulness of the diagram was obvious in the spatial task and designers spontaneously adopted it. It was not obvious in the temporal task. Performance on the two tasks became equivalent only when both groups were required to use the matrix diagram.

MATCHING DIAGRAMS TO PROBLEMS

A challenge for using diagrams to solve problems is to figure out which diagram to use for a particular problem. We have already seen a variety of options. Students in the design study used a *matrix* to design a floor plan and a manufacturing process. A matrix shows combinations of paired items such as the location of an office relative to the central corridor and to the accounting records. A *semantic network*, such as the one in Figure 1.2, shows how concepts are linked to other concepts. Some of the links represent *hierarchical* relations in which a large category (such as vehicle) is partitioned into smaller categories (car, boat). Chapter 4 ("Spatial Metaphors") discussed *Venn diagrams* to represent part–whole relations (Figure 4.2). Two neighbors, for example, may share ownership of a fence along their property line although they do not share the ownership of other structures.

Producing Diagrams

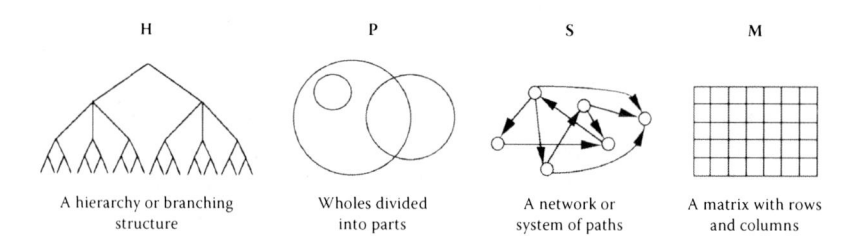

H	P	S	M
A hierarchy or branching structure	Wholes divided into parts	A network or system of paths	A matrix with rows and columns

FIGURE 8.5 A hierarchy, Venn diagram, network, and matrix. (From "Evidence for Abstract, Schematic Knowledge of Three Spatial Diagram Representations," by L. R. Novick, S. M. Hurley, and M. Francis, 1999, *Memory & Cognition, 27,* 288–308. Copyright 1999 by the Psychonomic Society. Reprinted with permission.)

Novick, Hurley, and Francis (1999) at Vanderbilt University became interested in whether college students have general knowledge about how to select an appropriate diagram for different problems. If students have general knowledge about using diagrams, asking them to generate problems that could be represented by a hierarchy, matrix, network, or Venn diagram would be a useful instructional exercise to prepare them for matching diagrams to problems. However, if students lack the knowledge to generate their own examples, then providing them with examples would be more helpful.

The instructions indicated that the purpose of the experiment was to determine which types of representations would be most beneficial for understanding particular problems. Students viewed the four diagrams in Figure 8.5 and either saw a problem that could be represented by each diagram (the specific example group) or generated their own problems (the general category group). During the test phase, students in the specific example group had to select which of the four *example problems* would provide the best representation for solving each test problem. Students in the general category group had to select which of the four *diagrams* in Figure 8.5 would provide the best representation for solving each test problem such as the Paraval Problem. If you are ambitious, you may want to go a step further and use your selected diagram to solve the problem. I have provided an answer at the end of this section.

The results showed that the general category group (58% correct) did better than the specific examples group (36% correct) on problems that could be represented by hierarchies, matrices, and networks. There was no difference between the two groups for problems that could be represented by Venn diagrams, presumably because there is more variation in the structure of Venn diagrams that make general knowledge less useful. The partial success of the undergraduates in selecting diagrams indicates that they have some general knowledge for the appropriate uses of hierarchies, matrices, and networks in solving problems.

In a follow-up study, Novick and Hurley (2001) proposed that there are 10 structural properties that can be used to distinguish between a matrix, a network, and a hierarchy. For example, one property is how one moves along a pathway.

There are no paths in a matrix, so it does not make sense to talk about moving along pathways. There is only a single path between concepts in a hierarchy because it is possible to move only up or down the links. There can be multiple paths between concepts in a network, although some paths typically have fewer links than others do. One property that should influence the choice of which diagram to use is whether no paths, one path, or multiple paths are needed to show the relations among the concepts.

Paraval Problem

A famous storyteller, recently returned from the land of Paraval, is recounting her travels there. The inhabitants of Paraval are most interesting and unusual. According to the storyteller,

> All Bandersnatches are Boojums.
> All Boojums are Snarks.
> Some Snarks are Frumious animals.
> No Snarks breakfast at five o'clock tea.

A young boy and his friends are fascinated by the tale. Afterward, they debate among themselves whether it is possible to conclude that

> (a) No Boojums breakfast at five o'clock tea.
> (b) No Frumious animals breakfast at five o'clock tea.
> (c) Some Snarks are Bandersnatches.
> (d) Some Frumious animals breakfast at five o'clock tea.
> (e) Some Boojums are Frumious animals.

Can you help the children answer these questions?

Novick and Hurley (2001) tested their proposed properties by asking college students to select and justify the type of diagram that would be most efficient for organizing information in each of 18 short scenarios. The college students in this study (computer science and mathematics education majors) were experienced users of diagrams. The results showed that these students not only did very well in selecting an appropriate diagram (with 88% accuracy when asked to choose between two diagrams), but their justifications mentioned 9 of the 10 properties proposed by Novick and Hurley as being important for making these decisions. Experienced users of diagrams know the criteria for selecting an appropriate one.

Figure 8.6 shows my Venn diagram of the Paraval Problem. The solid lines illustrate known relations and the broken lines illustrate unknown (but possible) relations. We can conclude that (a) no Boojums breakfast at five o'clock tea because these two sets do not overlap and that (c) some Snarks are

Producing Diagrams

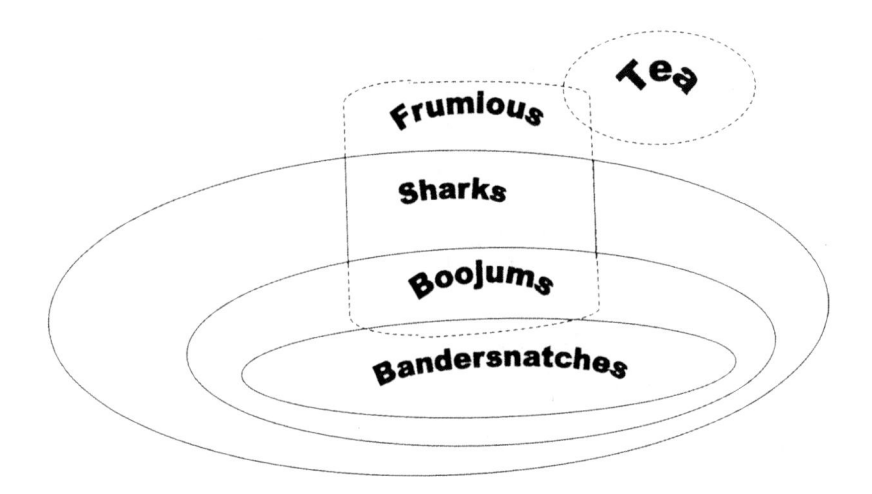

FIGURE 8.6 A Venn diagram representation of the Paraval Problem.

Bandersnatches because these two sets do overlap. We cannot make the other three conclusions. They may or may not be true, as indicated by the broken lines that relate these sets.

SPATIAL REPRESENTATIONS OF MEANING

We have seen how the effective use of diagrams helps students solve spatial problems, and have progressed from grade-school students to college students to college students who are experienced users of diagrams. Now it is time to look at the real experts in using diagrams—cognitive scientists. Okay, the real experts are architects, engineers, and builders but I have to write about what I know.

Diagrams are very helpful to cognitive scientists as they develop models of memory and thinking. The search space of the missionaries and cannibals problem shown in Figure 4.4 is one example and the semantic network shown in Figure 1.2 is another example. An important distinction made in Figure 1.2 is the contrast between amodal and perceptual symbols. Amodal symbols consist of abstract information such as a feature list or a semantic network, and perceptual symbols consist of reactivation of sensory images.

Although perceptual symbols form the basis for visual thinking, amodal symbols can also form the basis for visual thinking if we shift our attention from the people who are doing the thinking to the cognitive scientists who are modeling the thinking. Cognitive scientists who have developed semantic networks to model thinking have used diagrams to construct these models.

Figure 8.7 shows an example in which the ovals contain the concepts and the lines represent relational links between concepts. Some of these relations are hierarchical such as the line that connects *car* and *vehicle*, some specify features such as the line that connects *fire engine* and *red*, and some represent associations such as the line that connects *sunsets* and *sunrises*.

96   Thinking Visually

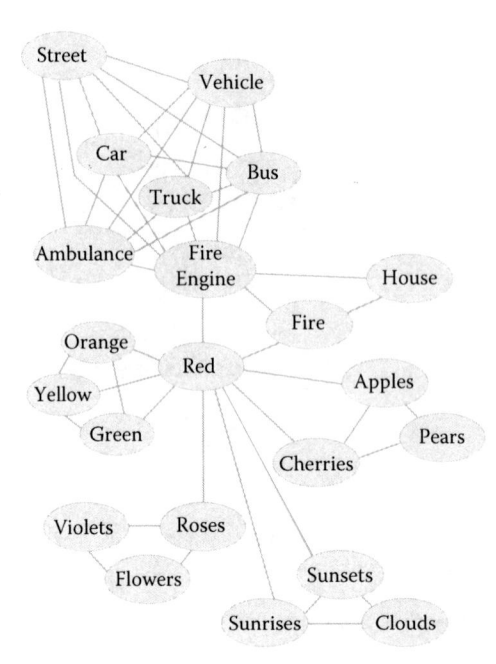

FIGURE 8.7 A semantic network model of conceptual organization. (From "A Spreading Activation Theory of Semantic Processing," by A. M. Collins and E. F. Loftus, 1975, *Psychological Review, 82*, 407–428. Copyright 1975 by the American Psychological Association. Reprinted with permission.)

The network in Figure 8.7 is part of a spreading activation model that was proposed by Collins and Loftus (1975). The model assumes that when a concept is activated, the activation unconsciously spreads along the links to activate other concepts. For instance, seeing an ambulance or the word "ambulance" would activate related concepts such as *fire engine* and *vehicle*. The degree of activation depends on the length of the link with stronger activation occurring for shorter links (*fire engine*) than for longer links (*vehicle*).

I am omitting the detailed assumptions and predictions of the spreading-activation model because it would take us off course in our examination of visual thinking. I include this model to make two points. The first is that the model uses a spatial diagram and a travel metaphor for thinking about amodal symbols. Thus, even amodal symbols rely extensively on visualization when we take a step back from the person doing the thinking to the cognitive scientist who is modeling the thinking. The second point, discussed in the next section, is that semantic networks are useful not only as a theory but as an instructional tool for organizing knowledge.

SEMANTIC NETWORKS AS INSTRUCTIONAL TOOLS

The effectiveness of semantic networks as instructional tools has been extensively investigated at Texas Christian University. In one study by Holley and Dansereau

Producing Diagrams

(1984), undergraduates received training on constructing semantic networks for material in their regular courses. These students and a control group of students then studied and were later tested on a 3000-word passage from a basic science textbook. Students who constructed semantic networks of this material did significantly better than the control group did on short-answer and essay questions.

These tests typically require a good organization of the major topics, and constructing semantic networks helps people organize knowledge. However, to improve their effectiveness, we need both instructional guidelines and software that enables students to construct large networks. The first need is being addressed by the research of O'Donnell, Dansereau, and Hall (2002) that forms the basis for the following recommendations:

- Begin with content that is extremely familiar to students so that they do not need to search for appropriate information.
- Use a number of well-constructed networks as initial examples.
- Include a discussion of the different types of links and the nature of the relationships among ideas.
- Ensure that students can recognize the corresponding relationships between information in the text and information in the networks.

The second need, effective computer software, is being addressed by the SemNet® software designed by my colleague, Kathleen Fisher. The process of constructing a network, according to its designers, is generally more valuable than the created network. The construction process is valuable because it

- engages students in thinking about the topic being learned
- requires them to transform their knowledge into new representations
- helps them see the relationship among ideas using categories, hierarchies, features, and causal links

In addition, the networks constructed by students provide insights to teachers about students' comprehension and understanding.

In conclusion, the previous chapter on viewing pictures and this chapter on constructing diagrams illustrate alternative ways of using visual displays. Pictures help us think concretely about objects and events. Diagrams help us organize knowledge. The compatible strengths of pictures for depicting particular cases and of semantic networks for organizing hundreds of cases can be combined in SemNet to link perceptual information with concepts organized in a semantic network. Clicking on the picture of the heart in Figure 8.8 would enlarge the picture to show labeled components. This is not only a productive use of an educational tool, in my opinion, but a potential theoretical model of how perceptual information can be embedded within large knowledge structures stored in long-term memory.

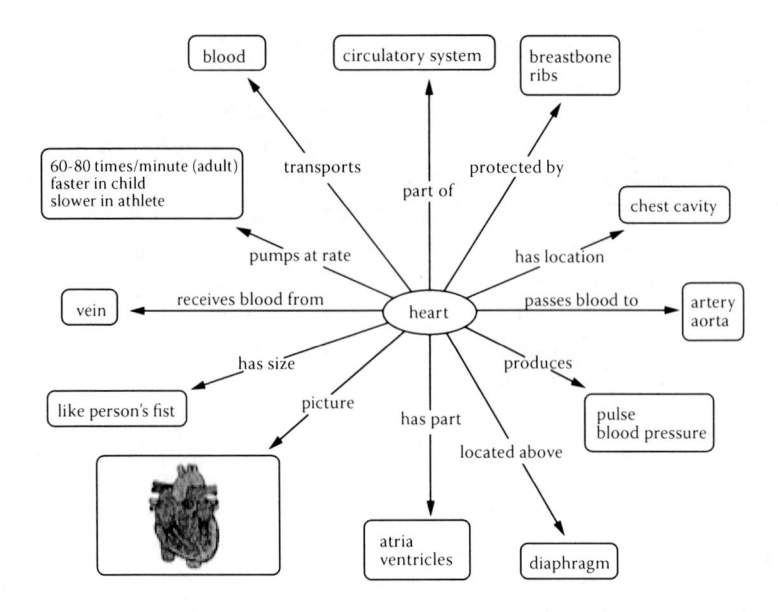

FIGURE 8.8 Part of a SemNet network centered on "heart." (From "Generating Connections and Learning in Biology," by M. Gorodetsky and K. M. Fisher, 1996. In K. M. Fisher and M. R. Kibby (Eds.), *Knowledge Acquisition, Organization, and Use in Biology* (pp. 135–154), Berlin: Springer. Copyright 1996 by Springer. Reprinted with permission.)

SUMMARY

Diagrams are abstract visual representations that preserve important spatial relations. They include matrices, networks, hierarchies, and Venn diagrams. The skillful use of these diagrams requires knowledge of how they differ in representing spatial relations. Psychologists, for example, have used networks to build models of semantic memory and educators have encouraged students to use networks to organize conceptual knowledge. Embedding pictures within networks combines the visual detail of pictures with the organizational advantages of diagrams. We will encounter more diagrams in Chapter 9 ("Comprehending Graphs"). The purpose of these diagrams is to display data.

9 Comprehending Graphs

Graphs are diagrams that display data. In the introduction to the second edition of his book, *The Visual Display of Quantitative Information*, Tufte (2001) informs us that each year between 900 billion and 2 trillion graphs are printed throughout the world. At their best, these graphs are instruments for reasoning about quantitative information.

I attended one of Tufte's talks several years ago. The audience sat at rows of tables and we were each given a 15 in. × 22 in. poster of Napoleon's march to Moscow during the War of 1812 (Figure 9.1). Tufte began his talk by explaining the poster. Beginning at the left side of the map, the thick upper band shows that Napoleon crossed the Polish-Russian border with an army of 422,000 men. The width of the band shows the size of the army at each point on the map. By the time the army reached Moscow, it had shrunk to 100,000 men. The lower dark band shows the retreat from Moscow, which had been sacked and deserted. It was a bitterly cold winter and many of the men froze to death as they retreated. The dramatic loss of life is vividly portrayed by the width of the two bands—the upper band showing the size of the army when it invaded Russia and the lower band showing the size of the army when it left Russia.

A French engineer named Charles Minard drew the graph of Napoleon's march in 1861. It displays four variables: (1) the size of the army, (2) its location on a two-dimensional map, (3) its direction of movement, and (4) (at the bottom) the temperature on various dates during the retreat from Moscow. Tufte believes that it may be the best quantitative graph ever drawn.

The graphs in this chapter are less elaborate than the one depicting Napoleon's march, but each illustrates how graphs demand and support visual thinking. They demand visual thinking because their message is usually not obvious from a quick glance. They support visual thinking because they can point to possible causal relations and important trends. Comprehending graphs requires understanding the graph and its implications.

REASONING FROM GRAPHS

Comprehending graphs has many important implications, including consequences for our personal well-being. An example (included in Tufte's [2001] book) is a graph that appeared in the 1964 Surgeon General's Report on *Smoking and Health*. The graph shows for various countries the relationship between the per capita consumption of cigarettes in 1930 and male death rate from lung cancer in 1950 (Figure 9.2).

Comprehending graphs involves cognitive skills such as reading the data, finding relationships in the data, and reading beyond the data according to Friel,

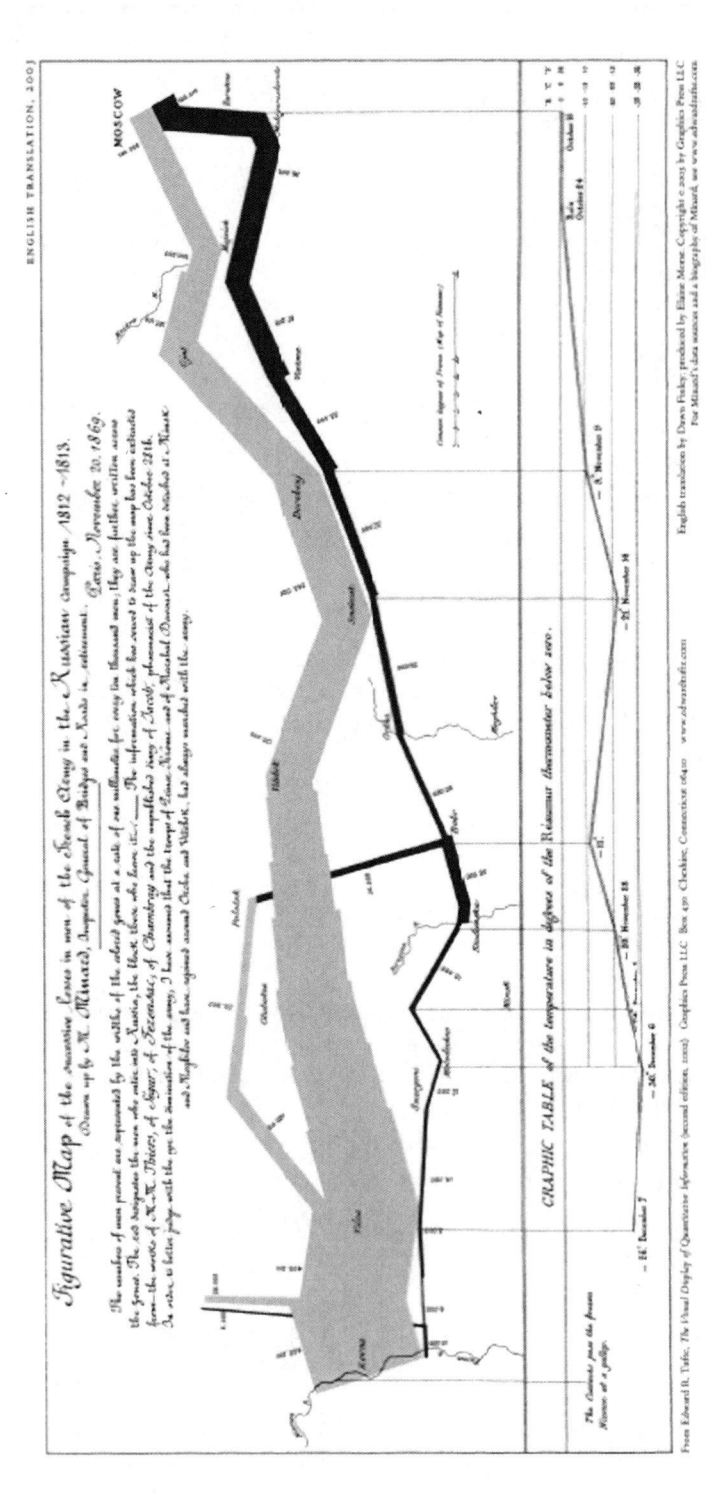

FIGURE 9.1 Napoleon's march to Moscow. (From *The Visual Display of Quantitative Information* (2nd ed.), by E. R. Tufte, 2001, Cheshire, CT: Graphics Press. Copyright 2001 by Graphics Press. Reprinted with permission.)

Comprehending Graphs 101

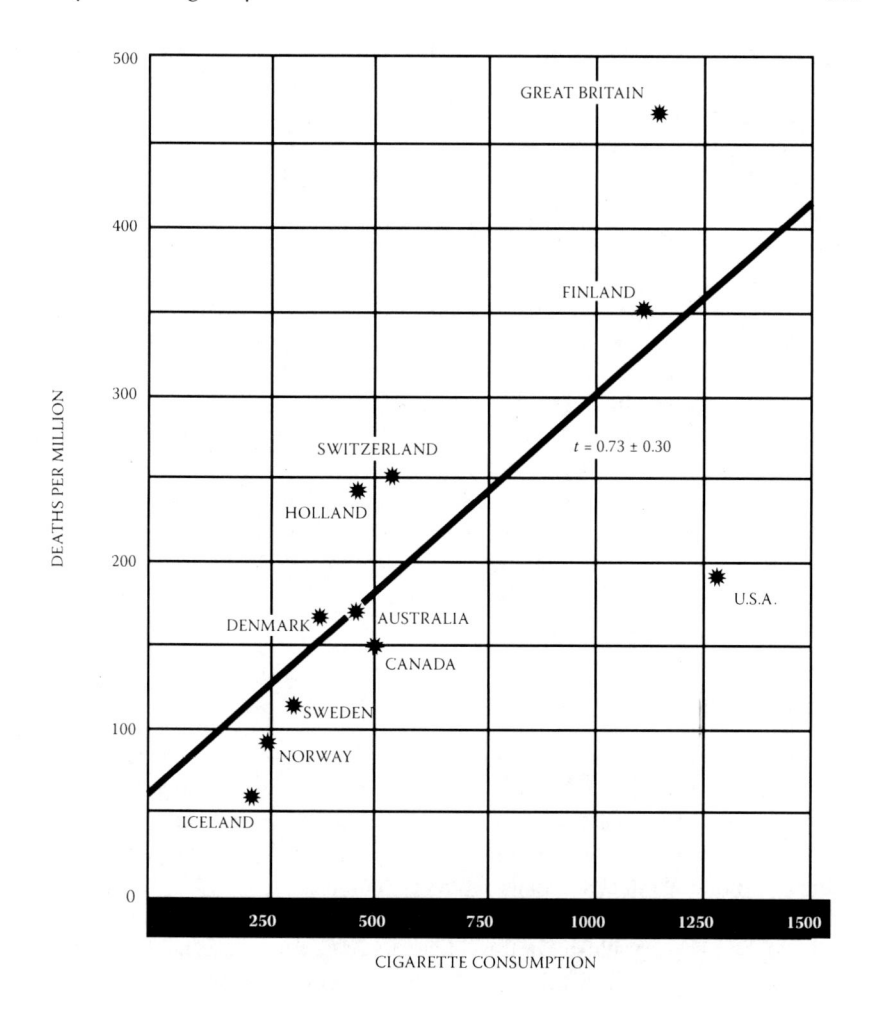

[Graph: Deaths per million vs. Cigarette consumption, showing data points for Iceland, Norway, Sweden, Canada, Australia, Denmark, Holland, Switzerland, Finland, Great Britain, U.S.A., with regression line labeled $t = 0.73 \pm 0.30$]

FIGURE 9.2 The relation between cigarette consumption and death rate from lung cancer. (Based on "Etiology of Lung Cancer," by R. Doll, 1955, *Advances in Cancer Research, 3*, 1–50. Copyright 1955 by Elsevier. Adapted with permission.)

Curcio, and Bright (2001) in their review of research on making sense of graphs. For example, I learned from *reading* the data in Figure 9.2 that cigarette consumption in Switzerland in 1930 was slightly more than 500 cigarettes per capita. I also learned that, in 1950, 250 males died from lung cancer for every million males living in Switzerland.

A second cognitive skill in comprehending graphs is *finding relations* in the data. The high correlation between cigarette consumption and deaths from lung cancer provides indirect evidence that there is a link between smoking and lung cancer. The evidence is indirect because a high correlation does not prove there is a causal relation between two variables. The data in Figure 9.2, however, do provide information that might help establish a causal link because the cause must precede the effect. A possible cause (cigarette consumption in 1930) precedes the effect (death from lung cancer in 1950) by 20 years because of the delayed effects of smoking.

A third cognitive skill is *reading beyond* the data. This can involve trying to explain data points that do not fit the overall pattern. Such points are called "outliers" as revealed for Great Britain and the United States. Which variables other than cigarette consumption might explain the relatively high death rates in Great Britain and the relatively low death rates in the United States? There are many possible candidates but one variable related to smoking is the amount of the cigarettes smoked. Although the number of cigarettes consumed was approximately the same in Great Britain and the United States, studies on discarded cigarette butts showed that American discards were significantly longer than British discards (Doll, 1955).

Figure 9.2 was one piece of an overwhelming amount of evidence presented in the 1964 Surgeon General's Report that linked smoking to lung cancer. I include the graph not because it, alone, provides a clear establishment for this link, but because it provides an example of how graphs can help us reason about important quantitative relationships between variables.

INTERPRETING EVENTS OVER TIME

The data displayed in Figure 9.2 show an association between two events that are separated by 20 years. However, graphs of many other events display how variables change continuously over time, such as the size of Napoleon's army and the temperature in Figure 9.1. One of the challenges in studying change over time is to understand how rates of change influence a quantity.

Figure 9.3 shows a graph that Kozhevnikov, Hegarty, and Mayer (2002) presented to their undergraduates at the University of California, Santa Barbara. The experimenters indicated that the figure graphed an object's motion and asked the

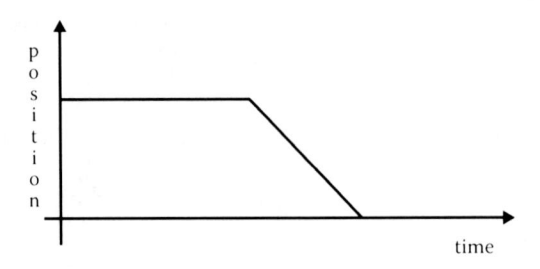

FIGURE 9.3 Students are asked to describe a real-world situation that corresponds to this graph. (From "Revising the Visualizer-Verbalizer Dimension: Evidence for Two Types of Visualizers," by M. Kozhevnikov, M. Hegarty, and R. W. Mayer, 2002, *Cognition and Instruction, 20*, 47–77. Copyright 2002 by Lawrence Erlbaum Associates. Reprinted with permission.)

Comprehending Graphs

viewers to describe a real situation that corresponded to the graph. I reviewed several studies in the previous chapter, including one by Hegarty and Kozhevnikov (1999), which showed some children incorrectly emphasized pictorial details rather than spatial relations when drawing diagrams. Kozhevnikov, Hegarty, and Mayer hypothesized that they would find similar individual differences in how college students interpret graphs.

Their findings supported their hypothesis. They interviewed 17 undergraduates who all strongly preferred visual processing to verbal processing. However, a battery of spatial ability tests revealed that 9 of these students were low-spatial visualizers and 8 were high-spatial visualizers. All 9 low-spatial visualizers interpreted Figure 9.3 as a picture in which the object travels down a hill. In contrast, the high-spatial visualizers correctly stated that the initial part of the graph indicates that the object does not move and the latter part shows that the object does move at a constant velocity.

The temperature graph in Figure 9.4 is a more challenging example that Carlson, Jacobs, Coe, Larsen, and Hsu (2002) presented to 20 high-performing, second-semester calculus students at Arizona State University. The graph shows the *rate* of change of temperature over an 8-hour period. Points above the horizontal axis show positive rates of change (increasing temperature) and points below the horizontal axis show negative rates of change (decreasing temperature). The task required constructing an approximate graph of temperature over the same period. In other words, it required translating rates of change into a graph of how the temperature changes.

Even though rate is a central concept in the study of calculus, Carlson and her colleagues found that most students were unable to construct a graph of temperature from the information presented in Figure 9.4. Only 4 of the 20 students drew an acceptable graph, and even those had difficulty providing a clear explanation of their graphs (Figure 9.5).

Given the graph of the rate of change of the temperature over an 8-hour time period, construct a rough sketch of the graph of the temperature over the 8-hour time period. Assume the temperature at time $t = 0$ is 0 degrees Celsius.

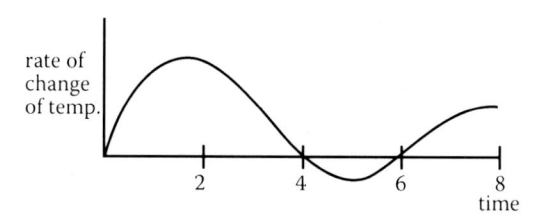

FIGURE 9.4 The temperature problem showing the rate of change of temperature. (From "Applying Covariational Reasoning While Modeling Dynamic Events: A Framework and a Study," by M. Carlson, S. Jacobs, E. Coe, S. Larsen, and E. Hsu, 2002, *Journal for Research in Mathematics Education, 33*, 352–378. Copyright 2002 by the National Council of Teachers of Mathematics. Reprinted with permission.)

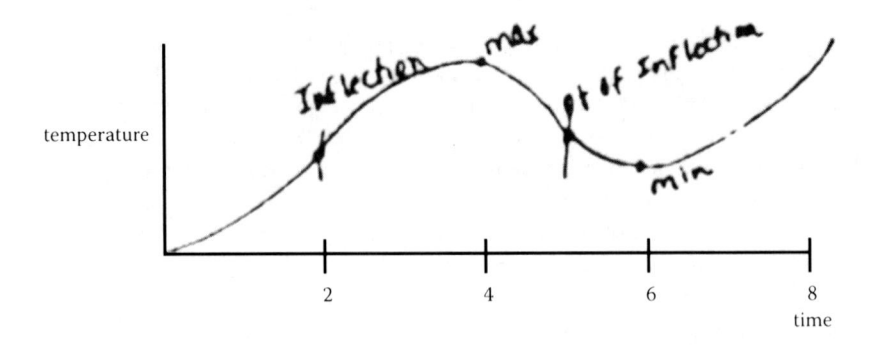

FIGURE 9.5 A calculus student's response to the temperature problem. (From "Applying Covariational Reasoning While Modeling Dynamic Events: A Framework and a Study," by M. Carlson, S. Jacobs, E. Coe, S. Larsen, and E. Hsu, 2002, *Journal for Research in Mathematics Education, 33*, 352–378. Copyright 2002 by the National Council of Teachers of Mathematics. Reprinted with permission.)

THE ALGEBRA SKETCHBOOK

The difficulty of linking rates of change (Figure 9.4) to changes in a quantity (Figure 9.5) motivated Yerushalmy (1997) at the University of Haifa to create a software environment called the Algebra Sketchbook. Her approach is to encourage students to view graphing as actions during major events so they can visualize how the graph changes during these events. The Pipeline Problem is an example.

The Pipeline Problem

A tanker loads fuel at sea. A single pipe first filled the tanker; then gradually a number of additional pipes are added. An announcement of an approaching storm leads to the pipes being disconnected one by one. After the storm, pipes that are again connected one by one as they were before the storm fill the tanker.

Figure 9.6 shows how the Algebra Sketchbook represents the pipeline situation. The events depicted in the graph are constructed from segments (icons) that show whether the rate of change is constant, increasing, or decreasing (Figure 9.7). The initial quantity of fuel added by a single pipe is represented by a straight line that shows an increasing function with a constant rate of change. The second event—adding more pipes—results in a curved function with an increasing rate of change. The third event—disconnecting pipes—results in a curved function with a decreasing rate of change. The amount of fuel in the tanker then remains constant during the storm, but begins increasing when pipes are again connected.

Comprehending Graphs

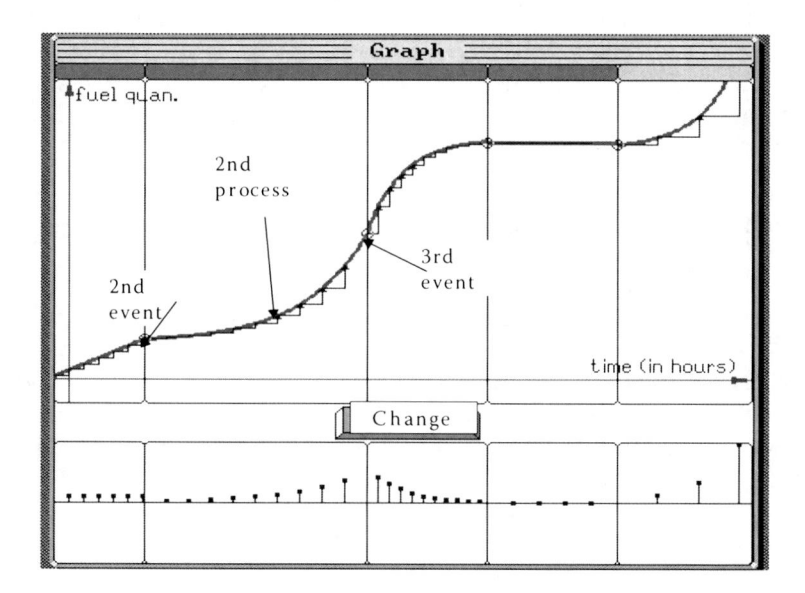

FIGURE 9.6 The Algebra Sketchbook's representation of the pipeline problem. (From "Mathematizing Verbal Descriptions of Situations: A Language to Support Modeling," by M. Yerushalmy, 1997, *Cognition and Instruction, 15*, 207–264. Copyright 1997 by Lawrence Erlbaum Associates. Reprinted with permission.)

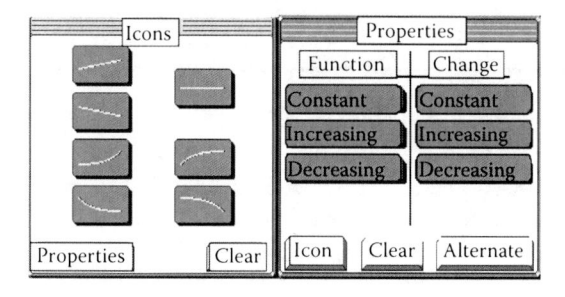

FIGURE 9.7 Properties of function pieces in the Algebra Sketchbook. "Mathematizing Verbal Descriptions of Situations: A Language to Support Modeling," by M. Yerushalmy, 1997, *Cognition and Instruction, 15*, 207–264. Copyright 1997 by Lawrence Erlbaum Associates. Reprinted with permission.

Yerushalmy (1997) studied the use of the Algebra Sketchbook by seventh-grade students in Israel who were beginning to learn algebra in an experimental curriculum called Visual Mathematics. The biggest challenge for these students, as we might expect from Carlson et al.'s (2002) research on rate of change, is to understand rate as a change in quantity over a standard unit of time. Many students instead used the iconic and verbal descriptions to represent rate as an independent concept rather than as a connection to a quantity. Although the students could use the icons to graph both rates and quantities, they had difficulty translating a graph of rates into a graph of quantities. I will return to this issue in Chapter 14 when I discuss the SimCalc software that dynamically links graphs to simulated events and to other graphs so students can receive visual feedback on the interrelation between graphs and events.

EXPONENTIAL GROWTH

The icons in the Algebra Sketchbook provide *qualitative* descriptions of change because they do not show the amount of change. There is no numerical scale along the y-axis. After students understand how quantities can change over time, they need to understand how quickly quantities change. This is particularly important when the rate of change involves exponential growth. This section explains why we need to understand exponential growth and the next section compares linear and exponential graphs as models of population growth.

The importance of understanding the consequences of exponential growth is emphasized by Cole (1998) in her book *The Universe and the Teacup: The Mathematics of Truth and Beauty*. She begins her chapter on exponential amplification by quoting Albert Bartlett, a former physics professor at the University of Colorado. According to Bartlett, the greatest shortcoming of the human race is our inability to understand the exponential function. He illustrates his claim with the Bacteria Problem. You will better understand exponential growth if you answer the questions posed in this section.

The Bacteria Problem

Two bacteria inhabit a Coke bottle at 11:00 am with the intention of beginning a new colony. Every minute the population doubles in size until it fills the entire bottle at 12:00 noon. At what time will the bottle be one-quarter filled?

The Bacteria Problem is solved by working backward. Because the population doubles in size every minute and is completely filled at 12:00 noon, it would then be half filled at 11:59 am and one-quarter filled at 11:58 am. Bartlett raises the question of when the bacteria should worry about overpopulation. Not at 11:58 am; the colony has filled only one-quarter of the bottle in its entire history! My own experiences support Bartlett's claim that people find it difficult to understand the explosive growth of the exponential function. Twenty-six of 45 students in a college algebra class at San Diego State University responded that the Coke bottle would be one-quarter filled at 11:15 am when one-quarter of the time had elapsed. Only 7 students wrote the correct answer, 11:58 am. I have obtained similar results from students in my cognitive psychology class.

Comprehending Graphs

Now answer the Salary Problem.

The Salary Problem

A movie star is offered a choice between two contracts for 20 days of filming. The first contract offers a flat rate of $100,000 for each day and the second contract offers one penny for the first day, with the daily salary tripling for each additional day of filming. Which should she chose?

After 20 days of filming, the star would receive $2 million at $100,000 per day. She would receive over $17 million if she initially received one penny and her salary tripled for each additional day of filming. Figure 9.8 shows a graph of how much the movie star would receive for three different methods of payment. The straight line shows how her money would grow if she received a flat rate of $100,000 per day. The two curves show what happens when money grows at an exponential rate. One curve shows the tripling of an initial salary of one penny. The other shows the doubling of an initial salary of $10. Notice that the initial accumulation of money is barely noticeable for these two payouts, but once some money accumulates it begins to skyrocket, as shown by the rectangular area of the graph. Now make estimates for the Pollution Problem.

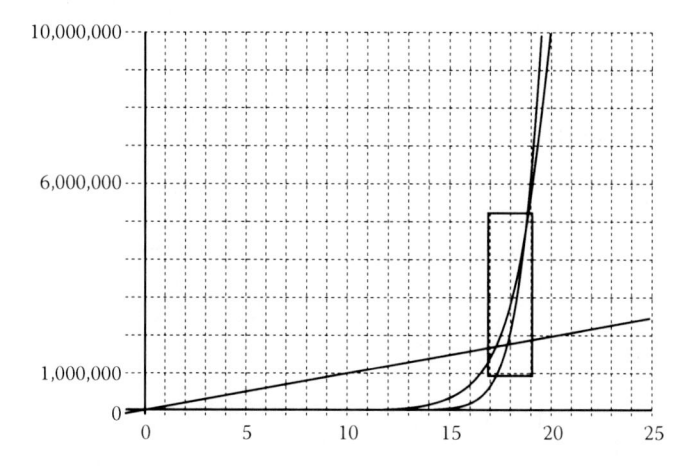

FIGURE 9.8 Graphs of linear and exponential growth. (From "Curricular Implications of Graphical Representations of Functions," by R. A. Philipp, W. O. Martin, and G. W. Richgels, 1993. In *Integrating Research on Graphical Representations of Functions,* by T. A. Romberg, E. Fennema, and T. P. Carpenter (Eds.), Mahwah, NJ: Lawrence Erlbaum Associates. Copyright 1993 by Lawrence Erlbaum Associates. Reprinted with permission.)

The Pollution Problem

A pollution index had the following increase in values for the years between 1970 and 1974: 3, 7, 20, 55, 148. How large will the pollution index be in 1979 if it continues the same rate of growth? Which year will the pollution index surpass 25,000?

The pollution index would surpass a value of 25,000 by 1979 if it kept growing at the exponential rate shown by the initial numbers. When college students at Pennsylvania State University were asked the question about the value of the pollution index in 1979, two-thirds produced estimates that were at or below 10% of the value of 25,000 obtained from exponential extrapolation. Another group was asked in which year the pollution index would surpass 25,000. Half of the students expected that a value of this magnitude would not be reached before the year 2000. Neither group could predict the explosiveness of exponential growth, as illustrated in Figure 9.8 for the Salary Problem.

Our inability to comprehend exponential growth is part of our general "number blindness" that has real-world implications. According to Cole:

> If we can't readily grasp the real difference between a thousand, a million, a billion, a trillion, how can we rationally discuss budget priorities? We can't understand how tiny changes in survival rates can lead to extinction of species, how AIDS spreads so quickly, or how small changes in interest rates can make prices soar. We can't understand the smallness of subatomic particles or the vastness of interstellar space. We haven't a clue how to judge increases in population, firepower of weapons, energy consumption. (Cole, 1998, p. 18)

MODELING POPULATION GROWTH

One of the possible real-world consequences of exponential amplification mentioned by Cole (1998) is population growth. The Animation Tutor™ includes a module (Population Growth) that provides users with an opportunity to evaluate the success of linear and exponential models of population growth in the United States during the 19th and 20th centuries. I selected this topic because understanding population growth is important, particularly when there is exponential growth.

A linear model of population growth in the United States during the 19th century is described by the following equation:

$$\text{Population} = \text{Population}_0 + \text{Rate} \times \text{Time}$$

where
Population is the population size at a given Time
Population_0 is the population size at the beginning of the period (1800)

Comprehending Graphs

Rate is the gain in population each year
Time is the number of years beyond the beginning year (1800)

You may recall that the equation for a straight line has a y-intercept and a slope. The y-intercept specifies the size of the population at the beginning of the century (Population$_0$) and the slope (Rate) specifies how many people are added each year.

The Population Growth module enables users to adjust the y-intercept and slope of the line to move the line close to the data points, as illustrated in Figure 9.9. This student's initial estimate was that the y-intercept was 5.0 million people and the slope was 0.7 million people (added each year). The student was satisfied with his initial estimate of the y-intercept, but adjusted the slope on his next two attempts to move the line closer to the data points. The goodness-of-fit (GOF) value of the line to the data points improved after decreasing the slope. A perfect fit would give a value of 1.000, but this is impossible to achieve unless all the points lie on a line. The best-fitting line to the data in Figure 9.9 results in a value of .924.

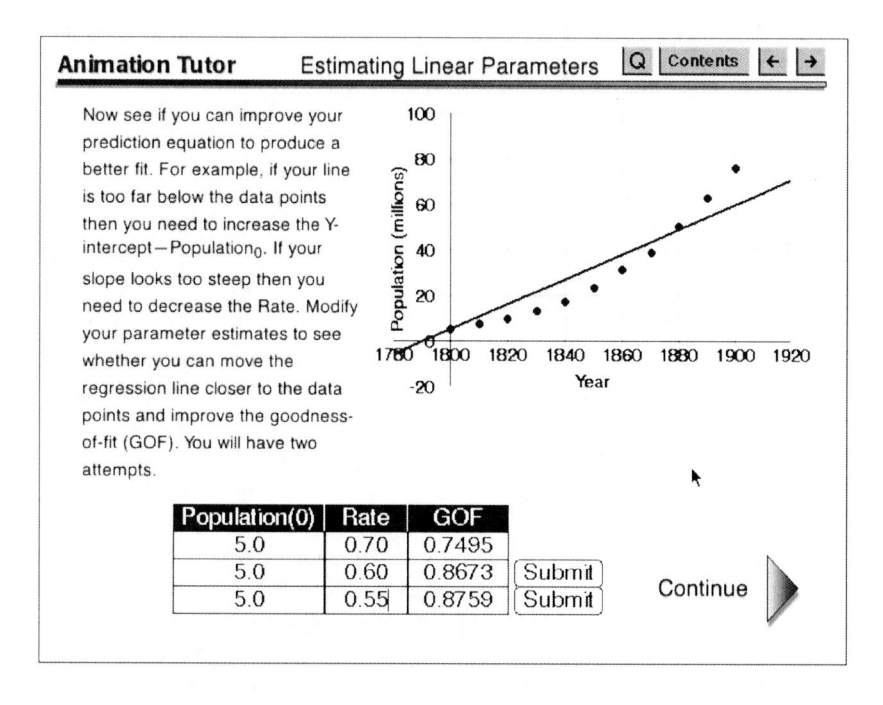

FIGURE 9.9 Example of successive attempts to improve the fit of a linear model in the Animation Tutor: Population Growth Module. (From Animation Tutor: Population Growth [Computer software], by S. K. Reed, B. Hoffman, and D. Short, D., 2009, San Diego, CA: San Diego State University. Copyright 2009 by San Diego State University. Reprinted with permission.)

The curved pattern of population growth during the 19th century suggests that an exponential model would fit the pattern better than a linear model. The equation for the exponential model is:

$$\text{Population} = \text{Population}_0 \times e^{\text{Rate} \times \text{Time}}$$

Population$_0$ is again the y-intercept, but Rate is now defined as the *percentage* of people added each year rather than the *number* of people added each year. The Animation Tutor™ provides the following information on doubling time to help people estimate the Rate parameter:

Annual percent increase:	1.0	2.0	3.0	4.0	5.0	6.0	7.0
Doubling Time (years):	69	35	23	17	14	12	10

This information shows that it would take 69 years for the population (or other quantities such as money) to double in size if it grew at a rate of 1% per year. Exponential growth does not care whether it is doubling people or money, but we should. We would be disappointed if it took a lifetime to double our investment. We should be concerned if it took only one lifetime to double a large population.

Users of the Population Growth module can study exponential growth rate by adjusting the y-intercept and Rate parameters to fit an exponential curve to the data points. The feedback is identical to the feedback provided for the linear model— an exponential curve is plotted based on the estimated parameters, accompanied by a GOF value. The best-fitting exponential model produces an almost perfect fit (GOF = .995) of the data based on an annual growth rate of 2.7%.

It would often be an advantage if we could predict the future by simply extending (extrapolating) the best-fitting model, but extrapolations are risky. Do you think that extrapolation of the linear model or the exponential model would best predict population growth in the 20th century? I hypothesized that students would select the exponential model because it was the best-fitting model and because they would not understand the explosiveness of exponential growth when extrapolated into the future. Thirty of the 33 students in our evaluation of the Population Growth module did select the exponential model over the linear model.

Figure 9.10 shows the result of extrapolating the best-fitting exponential and linear models into the 20th century. The linear model predicts too few people because the *number* of people added each year increased. The exponential model predicts too many people because the *rate* of growth slowed down in the 20th century. Even a growth rate of 2.7% would have resulted in a population of over 1 billion people in the United States by the year 2000, as illustrated by the exponential function flying off the graph.

Although the growth rate declined, an exponential model fits 20th century growth in the United States as well as it does 19th century growth if a growth rate of 1.3% is used in the model. However, even this apparently slow growth rate doubles population within our lifetime. Understanding exponential growth

Comprehending Graphs

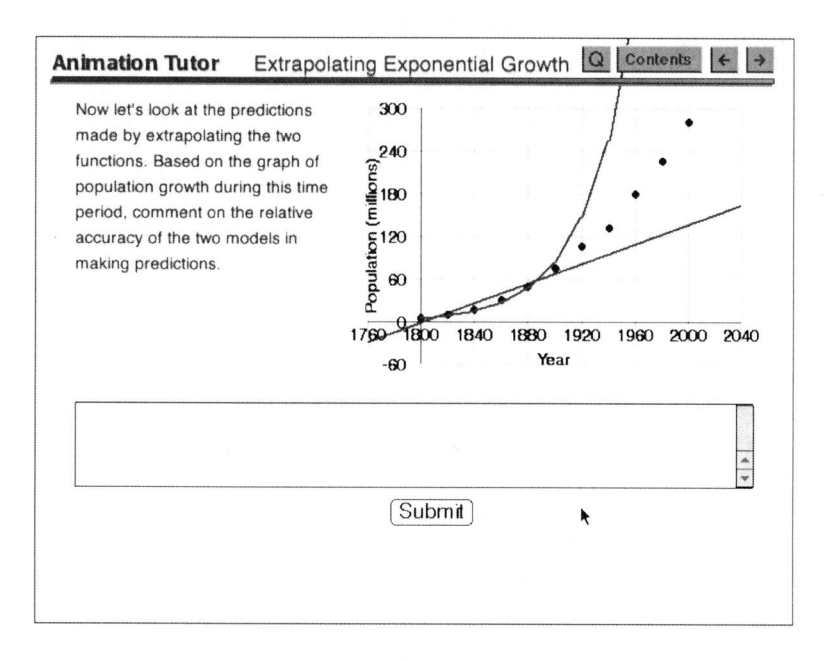

FIGURE 9.10 Extrapolating the optimal linear and exponential models to predict 20th century growth in the Animation Tutor: Population Growth Module. (From Animation Tutor: Population Growth [Computer software], by S. K. Reed, B. Hoffman, and D. Short, 2009, San Diego, CA: San Diego State University. Copyright 2009 by San Diego State University. Reprinted with permission.)

is particularly important when analyzing increases in the world's population. According to Joel Cohen (2000), the Abby Rockefeller Mauze professor of populations at Rockefeller University, it took 16 or 17 centuries to double the world's population to one-half billion; less than 200 years to double it again to 1 billion around the year 1800; about a century to double it again; and only 39 years for the last doubling. One-fifth of all the people who have lived in the last 6000 years are living today. As I stated in this chapter's introduction, comprehending graphs requires both understanding the graph *and* its implications.

SUMMARY

Comprehending graphs requires skills in reading data, finding relationships in data, and drawing conclusions from data. Understanding rates of change is particularly challenging. Software programs such as the Algebra Sketchpad enable us to represent changes in a quantity by combining linear and curvilinear segments. Understanding the implications of exponential growth, including population growth, is particularly important. Unfortunately, people typically underestimate its explosiveness and are therefore unable to understand its consequences.

10 Words and Pictures

Thinking visually does not often occur in isolation of other activities such as reading and acting. Instructions to assemble a toy or a piece of furniture, for instance, typically contain both words and pictures that we mentally combine to complete the task successfully. We use both to perform actions that combine the parts of the object.

Occasionally we have a wonderful opportunity to view the words and pictures of a creative writer and artist. After our family moved to La Jolla, California, in the fall of 1988, I decided to keep a look out for La Jolla's most famous resident, Theodore Geisel. Now if that name sounds unfamiliar, let me try his pen name—Dr. Seuss. At the time of his death in September 1991, Theodore Geisel's books had sold over 200 million copies. His success was the result of two great talents: writing and drawing.

The rest of us are typically found on the consumer end of the production line. We are the readers and viewers, rather than the writers and artists. However, we must still integrate words and pictures to enjoy and learn from both. Surprisingly, cognitive scientists know very little about how this integration occurs. They have painted their models with broad strokes, but have omitted the details. We hope that the broad strokes will lead us to the details, so let us take inventory of current progress, beginning with working memory.

INTEGRATION IN WORKING MEMORY

I mentioned in Chapter 4 that short-term memory (STM) is a temporary memory that is limited in both capacity and duration. Nonetheless, it is necessary for carrying out numerous tasks and has rightfully earned the title "working memory" for its important contribution to performing these tasks. Psychologists' initial conception of STM was that it could hold information indefinitely through verbal rehearsal. If you look up a number in the phone book, you can maintain it in STM through verbal repetition until you no longer need it. However, if you are performing a spatial task such as playing chess, you may want to store a visual image rather than a verbal description of the current game situation.

Baddeley and Hitch (1974) realized that working memory would need to store both words and pictures when they proposed their working memory model in 1974. Figure 10.1a shows that this model consisted of a visuospatial sketchpad for storing visual and spatial information, a phonological loop for rehearsing verbal information, and a central executive for managing the whole affair. Among the duties of the central executive is deciding how to utilize the verbal and spatial information for carrying out different tasks.

One such task requires chess players to reproduce the locations of pieces on a chessboard after briefly viewing the board. As chess players of various abilities

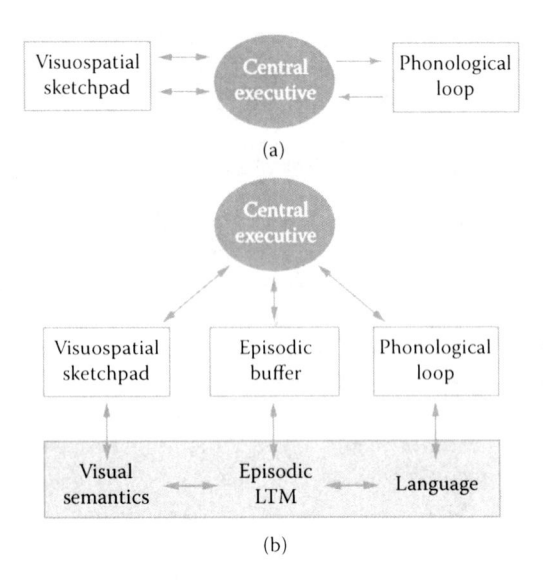

FIGURE 10.1 Baddeley's initial (a) and revised (b) model of working memory. (From "Is Working Memory Still Working?" by A. D. Baddeley, 2001, *American Psychologist*, 56, 851–864. Copyright 2001 by the American Psychological Association. Reprinted with permission.)

attempted to reproduce the board, they performed a secondary task that was designed to limit the use of either the phonological loop or the visuo-spatial sketchpad. The results of the study showed that suppressing speech had no effect but suppressing visual/spatial processing caused a marked impairment. These findings suggest that verbal coding does not play an important role in this task, but forming visual images is needed to have a good memory for the location of the pieces. Other research has confirmed that simply counting the number of pieces on the board or making decisions about moves is affected by tasks that interfere with visual/spatial processing but is unaffected by tasks that prevent verbal coding (Saariluoma, 1992).

It is remarkable that the working memory model proposed in 1974 went virtually unchanged for 26 years. Then in 2000, Baddeley proposed a modification of the model that is shown in Figure 10.1b. The two major changes were the addition of the episodic buffer to place more emphasis on the integration of visual and verbal information and the addition of long-term memory (LTM) to place more emphasis on the interplay between STM and LTM.

The problem with the original model was that there is no place to combine the visual and phonological information. In fact, most research on working memory was directed at identifying the independent contributions of space and speech in reasoning. Pictures and words must sometimes work together in harmony, however, and the episodic buffer provides a place for this integration.

An illustration of the difficulty of achieving such integration occurs in a cartoon that I have seen reproduced in several books. One man is giving another man directions. A thought balloon above the speaker shows a clear image of a street map that the speaker is trying to translate into words. A thought balloon above the listener shows a scrambled image of a street map that he has constructed from the verbal description.

Cognitive science currently lacks a detailed theory of how this integration occurs, or fails to occur. The scrambled map of the listener better portrays our understanding than does the coherent map of the speaker. However, we now recognize the problem and have a "box" (episodic buffer) where the integration occurs. We have already seen how container metaphors can be very useful and I expect that this new box will stimulate research to enhance our understanding.

COGNITIVE LOAD THEORY

The lack of a deep theoretical understanding of integration has not prevented cognitive scientists from discovering some useful instructional principles that can enhance learning of pictures and words. One very influential theory developed by John Sweller (2003) and his colleagues at the University of New South Wales examines the implications of a working memory that is limited in the amount of information that it can hold. According to Sweller's cognitive load theory, our efforts to combine information can result in cognitive overload—too much information for our working memory to manage.

Imagine for a moment that you have been assigned the design of an instructional manual to teach technicians how to test electrical appliances to ensure that the appliances are safe. The instruction must explain how to perform four tests on an electric kettle using a meter that measures resistance. You would most likely carefully write down the steps so the technicians could read each step and then perform it on the kettle and meter. The problem with this standard instructional approach is that it produces cognitive overload because of a *split-attention effect*.

The split attention effect occurs when people have to divide their attention between two sources, such as the instructions and the physical objects. The key to avoiding the split attention effect is to physically integrate the information, as shown in the illustration in Figure 10.2. Notice that the steps are numbered and placed next to the appropriate part of the diagram. The reader does not have to continuously switch attention between the instructions and the apparatus.

Sweller and Chandler (1994) evaluated the effectiveness of the modified instructions in an experiment with first-year trade apprentices from a company in Sydney, Australia. One group received standard instructions, which they applied directly to the physical apparatus. A second group received modified instructions consisting of variations of the diagram and steps shown in Figure 10.2. The physical integration of the instructional steps and diagram reduced the instructional time from 6 minutes for the standard instruction to 4 minutes for the modified instruction.

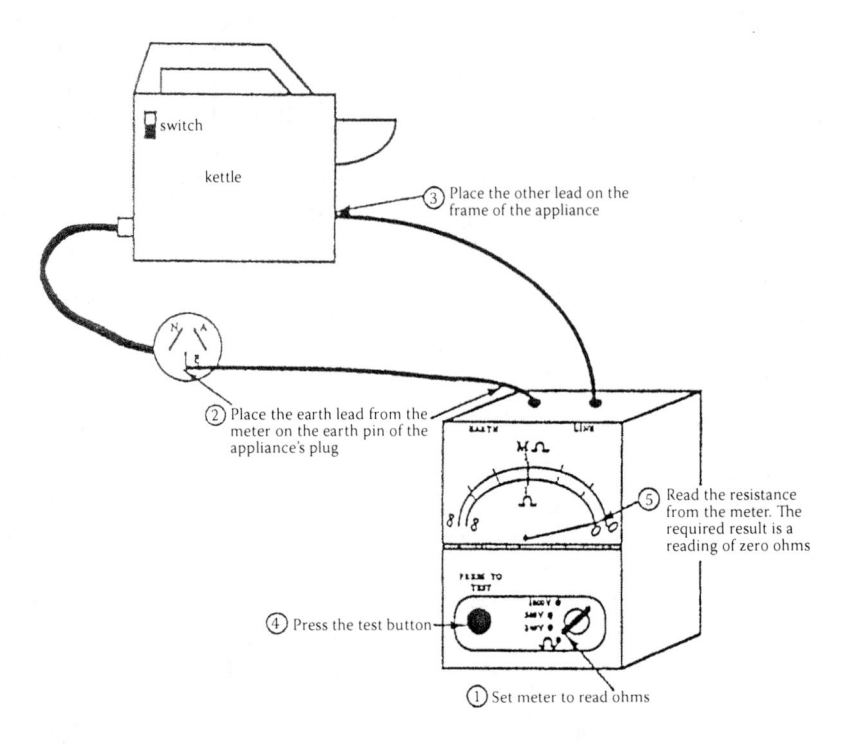

FIGURE 10.2 Modified instructions that physically integrate a picture and words. (From "Why Some Material is Difficult to Learn," by J. Sweller and P. Chandler, 1994, *Cognition and Instruction, 12*, 185–233. Copyright 1994 by Lawrence Erlbaum Associates. Reprinted with permission.)

More important, the modified instruction was much more effective. Technicians who received the modified instructions outperformed the standard group on both a written test and a practical test that included working with the physical apparatus. The better performance on the practical test was particularly impressive because the group who studied the modified instructions did not have any contact with the electrical apparatus until they did the practical test.

One could easily argue that an even better way to instruct the technicians would be to provide them with the modified instructions *and* the physical apparatus. Wouldn't they then have the best of both worlds? However, Sweller and Chandler (1994) proposed that these technicians would again suffer cognitive overload because of the *redundancy effect*. The redundancy effect states that if equivalent information is provided twice, then the added information simply provides more (rather than new) information. If interacting with the physical apparatus adds nothing to the information shown in Figure 10.2, then it is simply excess information that requires more time. Sweller and Chandler also included this condition in their experimental design and found their predicted results. The

Words and Pictures

group who had both the modified instructions and the physical apparatus did not perform any better than the group who had the standard instructions and the physical apparatus.

The split-attention and redundancy effects are two of eleven effects that Sweller (2003) discusses in a chapter on cognitive load theory. These effects are some of the best results that we have to help instructional designers benefit from cognitive theory. Let me conclude with one more effect that is relevant to all instruction, including the integration of pictures and words. According to the *expertise reversal effect*, instruction that reduces cognitive load for the novice may increase cognitive load for the expert. Adding text to a diagram can reduce cognitive load for the novice, but adding text to a diagram will increase cognitive load for the expert if the expert can understand the diagram without the text. The expertise reversal effect, the redundancy effect, and the split-attention effect are all supported by extensive research.

MAYER'S MULTIMEDIA THEORY

Another research program that has been supported by many experiments is the work on multimedia learning that Richard Mayer and his students have conducted at the University of California, Santa Barbara. Mayer's (2001) multimedia theory builds on Sweller's cognitive load theory, Baddeley's working memory model, and Paivio's dual coding theory that I discussed in Chapter 5. Mayer uses Paivio's proposal that information can be stored in either a verbal or a visual code, Baddeley's proposal that the integration of these two codes occurs in a limited-capacity working memory, and Sweller's proposal that successful instruction depends on avoiding cognitive overload.

Mayer used these theoretical assumptions and his own research findings to construct the multimedia learning model shown in Figure 10.3. The model focuses on the integration of words and pictures, but the words can be either heard or read depending on the instructional design. Working memory enables us to convert visual images to sounds, which occurs when we subvocalize written words and when we form a verbal description of a picture. Working memory also enables us to convert sounds to visual images, which occurs when we form a

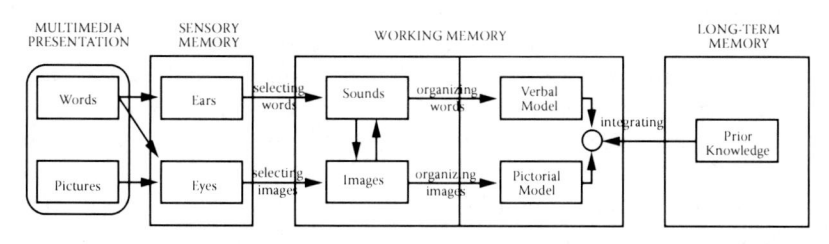

FIGURE 10.3 Mayer's multimedia model. (From *Multimedia Learning* (p. 44), by R. E. Mayer, 2001, Cambridge: Cambridge University Press. Copyright 2001 by Cambridge University Press. Reprinted with permission.)

mental picture from language. Integration of ideas occurs in working memory, as it does in Baddeley's working memory model. The sources of ideas are a verbal and a pictorial model constructed from the multimedia presentation and prior knowledge stored in LTM.

The dozens of experiments conducted by Mayer (2001) and his students have usually involved a science lesson (such as how a storm forms) or a description of some device (such as how a tire pump works). The instruction includes animations with either written or spoken text. Following the instruction, the students answer questions that measure both their retention of facts and their ability to apply these facts to novel situations. These experiments have resulted in seven principles for the design of multimedia instruction:

1. Multimedia principle: Students learn better from words and pictures than from words alone.
2. Spatial contiguity principle: Students learn better when corresponding words and pictures are presented near, rather than far from, each other on the page or screen.
3. Temporal contiguity principle: Students learn better when corresponding words and pictures are presented simultaneously rather than successively.
4. Coherence principle: Students learn better when extraneous words, pictures, and sounds are excluded.
5. Modality principle: Students learn better from animation and narration than from animation and on-screen text.
6. Redundancy principle: Students learn better from animation and narration than from animation, narration, and on-screen text.
7. Individual differences principle: Design effects are stronger for low-knowledge learners than for high-knowledge learners and for high-spatial learners than for low-spatial learners.

The first principle states that pictures are helpful, and I reviewed some of the benefits of pictures in Chapter 7. One exception noted in that chapter, based on the research of Harp and Mayer (1997), was that decorative pictures might not be helpful. This finding is generalized in the coherence principle, which states extraneous pictures, words, and sounds are counterproductive because they do not contribute to the message conveyed in the instruction. The effectiveness of pictures is also influenced by individual differences. Low spatial learners have the disadvantage that they must exert more cognitive effort to interpret the pictures, resulting in less cognitive capacity to integrate the pictures with the text.

Several other of Mayer's principles also follow directly from Sweller's cognitive load theory. The spatial contiguity principle corresponds to Sweller's split-attention effect—students learn better when words and pictures are near each other. The temporal contiguity principle generalizes this finding to time. An auditory explanation should accompany, rather than occur before or after, the picture.

Listening to an auditory explanation is preferable to reading an explanation because of the visual competition between reading words and perceiving pictures. Finally, the redundancy principle is an example of Sweller's redundancy effect. Reading the text while listening to a narration increases the amount of redundant information because the narration is a spoken version of the written text.

Mayer's principles should be very helpful for designing multimedia. One reservation, however, is that it is always wise to consider the specific instructional content when applying principles. Mayer's instruction explained the causal principles of lightning storms, tire pumps, and similar phenomena. In contrast, some material (such as formulas and equations) is difficult to describe verbally. My wife's uncle, a retired physicist, produced tape recordings of technical books for the blind. Describing equations was his biggest challenge. However, for people with sight, equations might serve the same role as pictures and be displayed visually on the screen while an audio narrative describes important features of the equation.

WORDS AND IMAGES

Mayer (2001) developed his multimedia model to study the integration of words and pictures. However, the model allows for the possibility that the reader or listener could convert words to visual images, as illustrated by the interconnection between sounds and images in Figure 10.3. Therefore, the pictorial model could be constructed from images rather than derived from a picture.

I presented preliminary evidence in Chapter 1 that readers construct images to help them understand text. You may recall the experiment by Stanfield and Zwaan (2001) in which participants read a sentence, pressed a bar when they understood it, and then judged whether a pictured object had been mentioned in the sentence. The findings supported the visual simulation hypothesis because the time to verify a mentioned object depended on whether the picture (such as a vertical or horizontal nail) matched the implied orientation in the sentence. The argument that visual simulations help us understand language is a particularly powerful one for supporting the importance of visual thinking. It says that understanding language itself rests on a foundation of visual thinking. I will therefore describe several experimental paradigms that converge on the conclusion that visual simulations support our understanding of language.

One paradigm is based on the concept of visual occlusions. Horton and Rapp (2003) argue that if visual simulations support text comprehension, then it should be more difficult to verify objects that disappear from sight. They illustrate an example from the classic story *The Strange Case of Dr. Jekyll and Mr. Hyde*:

> As the cab drew up before the address indicated, the fog lifted a little and showed him a dingy street, a gin palace, a low French eating-house, a shop for the retail of penny numbers and two penny salads, many ragged children huddled in the doorways, and many women of different nationalities passing out, key in hand, to have a morning glass; and the next moment the fog settled down again upon that part, as

brown as umber, and cut him off from his blackguardly surroundings. (Stevenson, 1895/1993, pp. 119–120)

The story allows us to experience the surroundings through the eyes of a visitor to the home of Mr. Hyde. However, these surroundings became occluded (hidden) by the fog, as they often do in stories in which doors close, people leave rooms, or objects move in front of people. Horton and Rapp (2003) proposed that if readers create images to understand narratives, the occluded objects should take longer to verify as having been mentioned in the story.

A typical story in their experiment described a patient named Marty who was in the hospital recovering from surgery. The story described the details of his room, including a tall vase of flowers by his bedside. In the occluded version of the story, a nurse draws a curtain around Marty's bed to give him privacy during an examination. In the nonoccluded version, the nurse attaches a monitor to his bed to take his blood pressure. A visual simulation of the story should continue to show the flowers only in the second version. At the end of the story, both groups were asked, "Did Marty have a vase of flowers?" The time to answer a question was longer for occluded objects, supporting the visual simulation hypothesis.

SIMULATED ACTIONS

My initial reaction to research on the verification of mentioned objects was that it provided convincing evidence for static images, but not for dynamic simulations that involve actions. Doesn't verifying that a nail or a vase of flowers was mentioned in the story require only a static image of the object? Then it occurred to me that the results of both studies likely depend on the simulation of actions. The quicker verification of either a horizontal or a vertical nail depended on whether the nail was pounded into the floor or the wall. The slower verification of the occluded vase of flowers depended on the nurse closing the curtain.

Glenberg and Kaschak (2002) at the University of Wisconsin have provided direct evidence that we mentally simulate actions when we comprehend text. In one of their experiments, college students had to quickly judge whether a phrase, such as "open the drawer" or "boil the air," made sense. They had to move their hand either toward their body or away from their body to hit a response key. Participants were faster in responding when the response required an action in the same direction as the one implied in the phrase. For example, they were faster in responding by moving their hand toward their body when verifying the statement "open the drawer" and faster when moving their hand away from their body for the statement "close the drawer." These findings are consistent with a mental simulation of the action that requires moving your arm toward your body to open a drawer and away from your body to close a drawer.

Thus far, we have been looking at how the mental simulation of actions helps skilled readers understand text. Do these results have any value for helping beginning readers? Would asking beginning readers to perform either physical or mental actions help them with comprehension and memory of text? Glenberg,

Gutierrez, Levin, Japuntich, and Kaschak (2004) investigated these questions by providing young readers with the opportunity to physically, and then mentally, manipulate objects described in a story.

Children in the second grade were initially shown commercially available toys that consisted of a farm scene (animals, tractor, barn, hay), a house with several rooms and people (mother, father, baby), or a garage scene (gas pumps, tow truck, car wash). The *Breakfast on the Farm* story reads as follows:

Ben needs to feed the animals.
He pushes the hay down the hole.
The goat eats the hay.
Ben gets eggs from the chicken.
He put the eggs in the cart.
He gives the pumpkins to the pig.
All the animals are happy now.

The experimenter began by naming each object for all of the children (Figure 10.4). The children in the manipulation condition then either physically manipulated the toys or imagined manipulating the toys after reading each sentence. Both actual manipulation and imagined manipulation greatly increased memory and comprehension of the text when compared to a control group that read the text twice without manipulation.

The researchers suggest two reasons why manipulation is helpful. The first is that manipulation helps young readers map the words to the objects they represent.

FIGURE 10.4 Farm toys. (From "What Brains Are For: Action, Meaning, and Reading Comprehension," by A. M. Glenberg, B. Jaworski, M. Rischal, and J. R. Levin. In *Reading Comprehension Strategies: Theories, Interventions, and Technologies,* by D. McNamara (Ed.), 2007, Mahwah, NJ: Lawrence Erlbaum Associates. Copyright 2007 by Lawrence Erlbaum Associates. Reprinted with permission.)

The second is that manipulation helps children derive inferences by constructing a mental model of the situation described in the story. Children who mentally manipulated the objects did better on spatial inference questions such as (1) "At the beginning of the story, is Ben on the same floor as the goat?" and (2) "Was the cart next to Ben when he got the eggs?" This information was not explicitly stated but could be inferred from the story.

INFERENCES

The last two questions illustrate that much of text comprehension requires inferring information that is not explicitly stated. We would likely infer, for example, that Ben is not on the same floor as the goat because he pushes hay down the hole. Inferences vary tremendously in difficulty. They also vary in type. One type is called a *predictive inference*. The Galloping Horse Problem requires making a predictive inference to which we will return later.

The Galloping Horse Problem

You are on a horse, galloping at a constant speed. On your right side is a sharp drop-off, and on your left side is an elephant traveling at the same speed as you. Directly in front of you is a galloping kangaroo and your horse is unable to overtake it. Behind you is a lion running at the same speed as you and the kangaroo. What must you do to get out of this highly dangerous situation?

Predictive inferences require the reader to predict what will happen next. Adult readers typically use visual simulations to make predictive inferences, as discovered by Fincher-Kiefer and D'Agostino (2004). Their Gettysburg College students read short scenarios such as:

The salesperson was sitting in the dining car of a train.
The waiter brought him a bowl of soup.
Suddenly, the train screeched to a halt.

The readers' spontaneous visual simulation of this text resulted in their making a predictive inference that the soup spilled.

However, visual simulations were not used to make a different kind of inference called a *bridging inference*, illustrated by the following text:

The salesperson was sitting in the dining car of a train.
The waiter brought him a bowl of soup.
Suddenly, the train slowed to a halt.
He jumped up and wiped off his pants.

The bridging inference is needed to connect the last sentence to the preceding sentences. It is likely that the salesperson wiped off his pants because the soup spilled, although we would not have predicted the spilling because the train slowed to a halt.

Bridging inferences look backward in time and depend on language according to Fincher-Kiefer and D'Agostino's (2004) results. Predictive inferences look forward in time and depend on visual simulations, as do inferences that establish causal relations. Research by Jahn (2004) at the University of Regensburg in Germany suggests that the spontaneous construction of spatial models is more likely when one event is causally related to another. For example, an approaching predator has greater causal relevance than an approaching nonpredator. Spatial constructions were more likely to occur for the sentence "Two zebras graze next to a shrub and a lion trots toward them" than for a sentence in which the phrase "a lion" is replaced by the phrase "an antelope." This example brings us back to the problem in which a lion chases a galloping horse. The answer to the problem is to get off the merry-go-round.

SUMMARY

Much of learning depends on our ability to integrate words and pictures. The integration occurs in working memory, which is limited in the amount of information that it can hold. Effective instruction must therefore prevent cognitive overload by following general principles such as those proposed by Sweller (2003) and Mayer (2001). Integration also occurs when readers create visual simulations of text. A variety of research paradigms have recently provided evidence that visual simulations support comprehension of spatial relations in text, including the generation of predictive and casual inferences.

11 Vision and Action

The integration of pictures and images with text is not the only occasion when visual thinking joins a team of skills. Vision and action form another team, as expressed by the term "eye–hand coordination." Ted Williams claimed that he could follow the flight of a pitched ball until it made contact with his bat. When later informed that this was impossible, he replied that it *seemed* like he could do it. I am sure it did—for a Ted Williams. You have read in this book about other situations in which vision and action join together to produce images that simulate action. Jack Nicklaus claimed that he never hit a shot, not even in practice, without having a very sharp picture of it in his head.

According to Koriat and Pearlman-Avnion (2003), much of what people remember concerns previously performed actions. When people become frustrated by memory failures, it is often forgetting to perform some task that elicits the frustration. In contrast to the importance of action in our everyday lives, much of the experimental research in psychology has emphasized the retention and retrieval of verbal material. Fortunately, psychologists have recently become more interested in studying memory for actions.

One way in which actions can improve memory is through creating an additional memory code based on motor movements. We saw in Chapter 5 during the discussion of Paivio's (1986) dual coding theory that two memory codes are better than one. Having both a verbal code and a visual code provides more opportunities to remember. Therefore, it should therefore not be surprising that a third memory code that preserves actions could also be helpful.

This third (motor) code came to my rescue early in my career after several years away from a typewriter. I am a lousy typist who typically makes several typos per sentence. I therefore decided to write the first and second drafts of my manuscripts by hand before giving them to a secretary to type. The arrival of computer-based word processors made it easier to correct typing errors so I decided again to do my own typing. I tried to recall the placement of the letters on the keyboard, first visually and then verbally, without success. I thought I would have to relearn this skill until I began to type. My fingers remembered the locations of the letters even though my verbal and visual memories had failed me.

MEMORY FOR ACTIONS

There is experimental support for the helpfulness of motor memories. Engelkamp (1998) reviews much of this early research in his book *Memory for Actions*. The typical experiment consisted of presenting participants with a list of 12 to 48 action phrases such as "nod your head" or "bend the wire" followed by free recall of the phrases. A verbal task consisted of simply listening to the phrases and a self-performed task consisted of acting out the phrases using either imaginary or actual objects. The superiority of enactment of the phrases was observed across many different experiments.

Koriat and Pearlman-Avnion's (2003) own research reveals why enactment aided recall. In their free recall task, people could recall the action phrases in any order. Evidence for the creation of *motor* codes would include recalling together phrases that had similar actions. A person who recalled the phrase "wax the car" might next recall "spread ointment on a wound" based on the similar motor movements. Evidence for the creation of *semantic* codes would include recalling together phrases that had similar meanings. A person who recalled the phrase "wax the car" might next recall "pour oil into the engine" because both phrases refer to cars.

Undergraduates at the University of Haifa were required to either enact each phrase (enactment instruction) or simply say the phrase aloud (verbal instruction). The enactment instructions required students to imagine the object and pantomime the described action as if the object were present. Notice how this method combines vision and action. Students in the enactment condition primarily recalled together phrases based on similar movements, whereas students in the verbal condition primarily recalled together phrases based on similar meanings. However, as found in many other experiments, the enactment of the phrases resulted in better recall than simply reading aloud the phrases.

Evidence from brain-imaging studies indicates the involvement of motor areas in the brain during the recall of enacted action phrases. In a study by Nilsson, Nyberg, Aberg, Persson, and Roland (2000), activity in the right motor cortex was strongest following encoding by enactment, intermediate following imaginary enactment, and lowest following verbal encoding. These motor memory codes provide additional opportunities for retrieval.

ACTING

The experimental support for the benefits of enactment primarily comes from traditional laboratory research in which undergraduates are asked to recall a list of action phrases. Do these results generalize to tasks that are more realistic?

Acting is an example of a more natural task that obviously requires action. I initially believed that an actor learns a script through verbal rehearsal in which she repeatedly reads the lines until they are memorized. However, research by Noice and Noice (2001, 2006) at Elmhurst College indicates that, for maximum effectiveness, a script should be learned through a process they call *active experiencing*, which treats emotion and movement as an important part of learning the dialogue.

Their initial insights came from interviews in which actors constantly referred to the relationship between learning lines and movement. In addition, months after the final performance of a play, actors could better recall dialogue that was accompanied by movement than dialogue delivered while standing in place. The Noices next studied whether undergraduates with no acting experience could better learn dialogue if they were encouraged to use the active-experiencing approach.

Students in three instructional groups worked in pairs to learn dialogue from the play, *The Dining Room*, in which two siblings argue over dividing their parents' furniture. Pairs of students in the full active-experiencing group previously learned motor movements before studying the dialogue so they could enact the movements when studying the script. They and the partial active-experiencing pairs (who sat on chairs) were then instructed to actively communicate the meaning of the lines to their partner rather than focus on memorizing the dialogue. In contrast, the memorization pairs were told to focus on memorizing the exact text, rather than acting the role. Figure 11.1 shows that the full active-experiencing approach resulted in the best recall of the dialogue.

These findings are consistent with the laboratory findings in which enactment improved recall of verbal phrases. However, the Noices discuss a number of differences between dialogue learning and the free-recall paradigms discussed previously. The previous results were based on recalling a list of action phrases rather than connected text. In addition, the actions literally enacted the action phrases whereas the actions in a play usually accompany the dialogue rather than act it out. An actor who crosses his legs does not say, "I think I will cross my legs now," but rather uses the action to provide a context for the dialogue. In spite of these differences, enactment of the script facilitates the recall of verbal information.

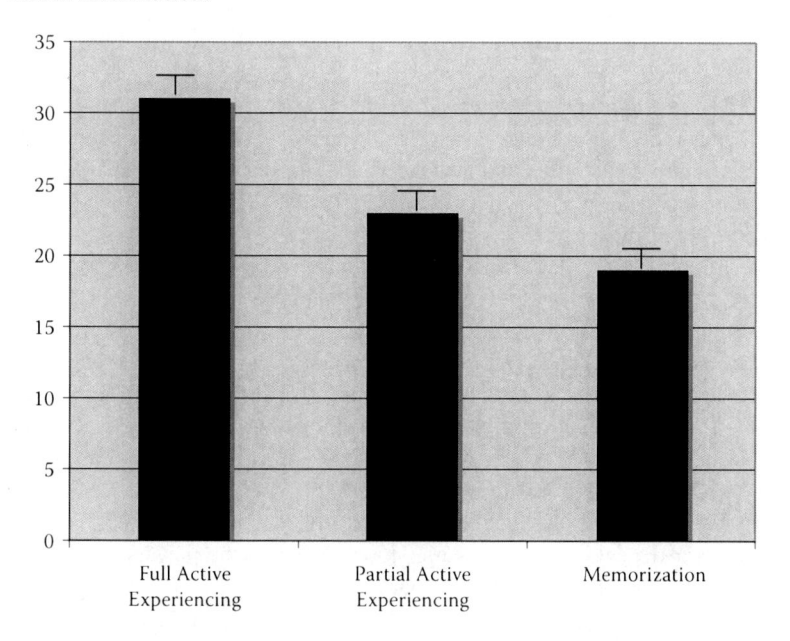

FIGURE 11.1 Percentages of narrative recalled for the full active-experiencing, partial active-experiencing, and memorization groups. (From "What Studies of Actors and Acting Can Tell Us About Memory and Cognitive Functioning," by H. Noice and T. Noice, 2006, *Current Directions in Psychological Science, 15*, 14–18. Copyright 2006 by Blackwell. Reprinted with permission.)

The Noices have subsequently generalized the active-experiencing approach to help older adults (aged 65 to 90 years) improve their performance on cognitive tasks (Noice, Noice, & Staines, 2004). They designed a 4-week course in which a professional theater director presented a series of progressively more difficult exercises to encourage the adults to become cognitively, emotionally, and physically involved in the performance. The participants were tested on problem solving and word recall exercises before and after taking the course, and dramatically improved their scores on both tests. The improvements were likely caused by the mental stimulation provided in the course because the class exercises were not specifically directed toward the test questions. In contrast, participants in an art appreciation course did not improve on the problem solving and word-recall tests.

CHESS

The initial examples show how action can aid people's memory for skills such as typing, recalling phrases, and learning dialogue. Now it is time to switch gears to consider not how action improves memory, but how observing people's actions can provide information about the organization of their memory. As shown in the previous chapter, chess players use visual memory to reproduce the pieces on a chessboard. Carefully observing how they replace the pieces provides important clues about their memories.

The classic research on chess was a study by the Dutch psychologist Adrian de Groot during the 1940s that was later published in his 1965 book *Thought and Choice in Chess*. De Groot was interested in why master chess players are so much better than less skilled players. He began by asking players to think aloud as they planned their moves and discovered that master chess players did not consider more moves or plan further ahead than the weaker players. Of course, they did select better moves.

De Groot (1965) hypothesized that the master players selected better moves because they *perceived* the board differently than the weaker players. Empirical support for de Groot's conclusion came from a series of clever experiments that required players of different abilities to view and then reproduce a chessboard as it might appear 20 moves into a game. Figure 11.2 shows two of the board configurations that de Groot used in his study. After a chess player viewed the board for 5 seconds, de Groot covered the board, removed the pieces, and asked the player to reproduce the locations of the pieces.

The five master players correctly located 90% of the pieces, compared with only 40% for the five weaker players. However, when the players viewed pieces that were placed randomly on the board, the master players no longer had an advantage over the weaker players. De Groot (1965) argued that the master players were better able to remember the locations only when they could organize the pieces into familiar groups.

Chase and Simon (1973) extended de Groot's paradigm to identify the familiar groups (called "chunks") that presumably caused the superior performance of the master players. They evaluated a master chess player, a class-A player,

Vision and Action 129

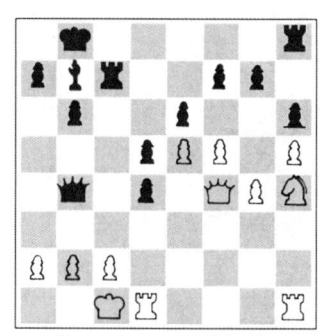

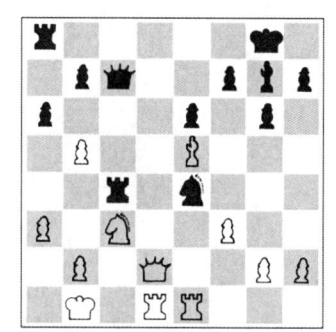

From: Janosevic-Krisnik;
Zenica 1964

From: Bannik-Geller
Moskou 1961

FIGURE 11.2 Examples of memorized chessboards. (From "Perception and Memory Versus Thought: Some Old Ideas and Recent Findings," by A. D. de Groot. In *Problem Solving: Research, Method, and Theory*, B. Kleinmuntz (Ed.), 1966, New York: Wiley. Copyright 1966 by Wiley. Reprinted with permission.)

and a beginner on de Groot's reproduction task and measured the time between the placements of successive pieces. Chase and Simon assumed that pieces belonging to the same chunk would be placed on the board as a group. They therefore classified pauses greater than 2 seconds as indicating chunk boundaries in which the players were planning the placements of the next group. The pauses suggested that the average number of chunks was 7.7 for the master player, 5.7 for the class-A player, and 5.3 for the beginner. The number of pieces per chunk averaged 2.5 for the master player, 2.1 for the class-A player, and 1.9 for the beginner. The players therefore differed in both the number and size of the chunks, supporting de Groot's (1965) hypothesis that the better players have a perceptual advantage over the weaker players. This perceptual advantage helps them select better moves and, in this case, helped them remember the location of the pieces.

REASONING

The combination of vision and action can also aid simpler reasoning tasks. Schwartz and Black (1999) did a study at Vanderbilt University that illustrates how action and visual imagery combine to help people make judgments about pouring water from a container. Look at Figure 11.3 and decide which of the two glasses you would have to tilt further in order to pour the water. On the other hand, would the tilt be the same for both glasses?

Schwartz and Black (1999) gave this task to 16 undergraduates by initially showing them two glasses in which a thin strip of tape indicated the water level. They were told to close their eyes and physically tilt each cup until the imaginary water reached the rim. Each cup was then taken from the tilters before they opened their eyes. All 16 correctly tilted the narrower cup further. After the tilting phase,

FIGURE 11.3 Will the water pour sooner from the narrow glass or the wide glass? (From "Designs for Knowledge Evolution: Towards a Prescriptive Theory of Integrating First- and Second-Hand Knowledge," by D. L. Schwartz, T. Martin, and N. J. Nasir. In *Cognition, Education, and Communication Technology* (pp. 21–54), P. Gardenfors and P. Johansson (Eds.), 2005, Mahwah, NJ: Lawrence Erlbaum Associates. Copyright 2005 by Lawrence Erlbaum Associates. Reprinted with permission.)

participants were explicitly asked, "Do you think the water pours out at the same or different angles for each cup?" If they replied "different," they had to indicate which would have to be tilted further. Only 3 of the 16 participants replied that the narrower cup would be tilted further. Six replied that the cups would be tilted the same and 7 replied that the wider cup would be tilted further.

These findings demonstrate that knowledge revealed through action can be more accurate than verbally expressed knowledge. Another experiment by Schwartz and Black (1999) demonstrated that students were fairly accurate when only imagining, rather than performing, the rotation of the glasses. However, anecdotal evidence suggested the lack of bodily motion made the imaging task more difficult. The tilters took more time and occasionally had to restart their mental simulations.

Bodily motion also plays an important role in conversation in the form of gestures. Gestures enable us to demonstrate concepts such as size that may be more difficult to state verbally. They also facilitate reasoning as shown by the finding that gesturing *reduced* the cognitive demands on working memory when students explained mathematical solutions. Wagner, Nusbaum, and Goldin-Meadow (2004) at the University of Chicago asked college-age adults to factor quadratic equations on a white board and then explain their solutions. To determine the memory demands of the explanation task, the researchers gave the students supplementary information, such as a random string of letters, before they explained their solution. Students were later able to recall more supplementary items if they gestured while explaining their solution (as shown in Figure 11.4). The number of recalled items depended on the meaning of the gestures, with more items recalled when the gestures and verbal explanations conveyed the same meaning. Gesturing therefore reduced the memory demands of explaining the solution, particularly when the gestures and verbal explanation were compatible.

You may wonder why the gestures and verbal explanations would not be compatible. The answer is that people occasionally say one thing and do something

Vision and Action 131

FIGURE 11.4 Gesturing while explaining a solution. (From *Hearing Gesture: How Our Hands Help Us Think* (p. 152), by S. Goldin-Meadow, 2003, Cambridge, MA: Harvard University Press. Copyright 2003 by Harvard University Press. Reprinted with permission.)

else. Although this can at times be frustrating, it can also be revealing. The mismatch between information conveyed by gesture and by speech provided useful diagnostic information, such as determining when students were considering different solution options as they reasoned about problems.

Consider a Piagetian task in which children are asked whether the two containers in Figure 11.3 contain the same amount of water. Many children would reply that the amount of water is the same because the height of the water is identical. Some children will give compatible gestures by pointing to the height of the water, but others will give incompatible gestures by showing an (unconscious) awareness of the different widths as they attempt to grasp the containers. Those children whose gestures show an understanding that is not revealed in their verbal responses are more likely to benefit from instruction. As pointed out by Goldin-Meadow (2003), these children appear to have the correct knowledge at their fingertips.

REFLECTIONS

Realizing that the two containers in Figure 11.3 hold different amounts of water requires both recognizing that the containers differ in diameter and realizing that the diameter influences the amount of water in the container. Piaget (1977) emphasizes this distinction between noticing and reflecting in his book *Recherches sur l'abstraction réfléchissante*.

Campbell (2001) translated this book into English with the title *Studies in Reflecting Abstraction*. He prefaces his translation with an essay on the fundamental ideas in Piaget's theories. According to Campbell:

> For Piaget, knowledge is not a matter of pictures or sentences or symbolic data structures in the mind that correspond to structures in the world; knowledge is basically pragmatic or action-oriented. Knowledge is fundamentally *operative*; it is knowledge of what to do with something under certain possible conditions. (Campbell, 2001, p. 3)

Campbell (2001) expands upon this foundation in describing other key concepts in Piaget's theory, particularly the concept of reflecting abstraction that did not even appear in Piaget's theory until 30 years into his research program. Reflecting abstraction is needed to modify the cognitive structures that represent operative knowledge.

Campbell (2001) cites an example from Chapter 2 of his translation in which Piaget uses poker chips to teach children the concept of multiplication as repeated addition. For instance, children are asked to place three chips in a row, followed by placing another three chips in the same row. According to Piaget, children have to perform two types of abstraction to think about multiplication as repeated addition ($2 \times 3 = 3 + 3$). The first is to recognize how many chips they are adding each time. This is an example of *empirical abstraction*, which is abstraction of information that children observe in the environment. The second is to keep track of the number of times that they add the same amount. This requires the use of *reflecting abstraction* to abstract a property of their actions. Reflecting abstraction is required to create new knowledge such as recognizing that adding two chips three times produces the same number of chips as adding three chips two times ($3 \times 2 = 2 \times 3$).

MANIPULATIVES

An instructional challenge raised by Piaget's emphasis on reflecting abstraction is to assure that actions are accompanied by reflections. For example, a student could perform actions on the poker chips or other physical objects (manipulatives) without understanding the connection to mathematics. Good manipulatives, according to Clements (1999), are those that are meaningful to the learner, provide control and flexibility, and assist the learner in making connections to cognitive and mathematical structures. His own instructional program called *Shapes* allows young children to use computer tools to move, combine, duplicate, and alter shapes to make designs and solve problems.

In commenting on the mixed results of research on the effectiveness of manipulatives in teaching mathematical concepts, Thompson (1994) argued that it is necessary to look at the total instructional environment to understand the effectiveness of concrete materials. Although the material may be concrete, the idea the material is supposed to convey may not be obvious because of students' ability to create multiple

Vision and Action 133

interpretations of materials. To draw maximum benefit from students' use of these materials, Thompson proposes that instructors must continually ask, "What do I want my students to understand?" rather than "What do I want my students to do?"

Moyer (2002) at George Mason University demonstrates the importance of looking at the total instructional environment. She tracked how 10 teachers used manipulatives after they attended a 2-week summer institute on the use of a Middle Grades Mathematics Kit that included base-10 blocks, color tiles, snap cubes, pattern blocks, fraction bars, and tangrams. The teachers made subtle distinctions between real math that used rules, procedures, and paper-and-pencil tasks and fun math that used the manipulatives. Unfortunately, the fun math typically was done at the end of the period or the end of the week and was not well integrated with the real math.

However, recent research inspired by theories of embodied cognition is discovering *when* manipulatives enhance learning. One example is the work of Arthur Glenburg and his colleagues (2004, 2007) that was described in the previous chapter. Glenberg has extended his research on the use of manipulatives during reading instruction to investigate the use of manipulatives during mathematics instruction. By using zoo and fair toys that are similar to the farm toys in Figure 10.4, Glenberg's group has been teaching third graders to solve story problems such as the following:

> There are two hippos and two alligators at the zoo.
> They live by each other, so Pete the zookeeper feeds them at the same time.
> It is time for Pete to feed the hippos and the alligators.
> He gives each hippo seven fish.
> He gives each alligator four fish.
> Hippos and alligators are happy now that they can eat.
> How many fish do both the hippos and the alligators have altogether before they eat any?

Children in the *story-relevant manipulation* group would solve this problem by counting out the appropriate number of little toy fish and distribute them to the animals. Children in the *abstract manipulation* group would count out Lego pieces to represent the fish without the presence of the story objects. Children in the *re-read* group would read each sentence aloud and reread the most relevant sentences. The third graders in the story-relevant manipulation group correctly solved 75% of the problems, compared to 28% and 35% for the children in the other two groups. In a second experiment, Glenberg's team showed that the children also did very well with imagined manipulation of the objects after they had learned how to solve the problems with physical manipulation a week earlier.

The effectiveness of the story-relevant manipulations is very impressive but the ineffectiveness of the abstract manipulations illustrates that there are constraints on the effective use of manipulatives. If the manipulatives have to depict objects in the story, then they will be difficult to use for problems such as the following:

Pete the zookeeper is in charge of feeding animals at the zoo.
In the morning, Pete feeds the polar bear and the monkey.
First, he feeds the polar bear nine fish.
Then he feeds the monkey five bananas.
In the afternoon, he feeds the polar bear eight fish.
Now, Pete remembers he has to feed the monkey again.
If the monkey gets the same number of things as the polar bear, how many bananas should Pete give the monkey in the afternoon?

Solving the problem by summing 9 and 8 and subtracting 5 creates some problems for the story-relevant manipulation group. The problem does not specify the final step and the child may have difficulty subtracting 5 bananas from 17 fish. Children in the story-relevant manipulation group correctly solved only 46% of these problems, compared to 60% for the other two groups. The instructional challenge will be to build on the success of concrete manipulatives for the simpler problems to promote thinking that is more abstract. We will encounter a successful example in Chapter 13.

SUMMARY

Vision and action combine to improve both memory and reasoning. Pantomiming verbal phrases increases recall relative to saying the phrases aloud. Actors recall dialogue better when it is accompanied by movement. This active-experiencing approach also improves the memory of students and older adults. Observing people's actions provides information about the organization of a chess player's memory and the existence of alternative strategies for solving math problems. As emphasized by Piaget, however, actions must be accompanied by reflective thinking to create new knowledge. Manipulating manipulatives in a manner that encourages reflection can increase the understanding of mathematics.

12 Virtual Reality

Both vision and action come together when we interact with a virtual environment. Virtual reality creates concrete experiences by producing physical simulations of the real world. According to Sue Cobb at the University of Nottingham and Danae Fraser at the University of Bath, there are no single concise or generally accepted definitions of virtual reality (Cobb & Fraser, 2005). However, their chapter on multimedia learning gives many examples of the application of virtual reality to classroom learning, rehabilitation, spatial cognition, and social skills training. Consequently, with this wide range of applications, and the multiple ways in which virtual environments can differ from real environments, psychologists are only beginning to understand how virtual environments support learning.

We are making progress, however, as indicated by the examples in this chapter. My plan is to begin by presenting some examples of virtual reality to illustrate the variety of ways it can simulate reality. I then discuss some broad theoretical efforts that can provide a framework for the development of more detailed theories for specific applications. Finally, I discuss several research projects to show the theoretical and practical value of virtual reality to study and improve perception and action in risky environments.

EXAMPLES OF VIRTUAL REALITY

Virtual environments can appear in surprising places, such as the artificial surfing environment that opened near the ocean in San Diego's Belmont Park. Producing artificial surf would not be surprising in Arizona, but next to the ocean? The justification is that the Pacific Ocean often lives up to its name (implying calmness) as surfers wait for big waves that never arrive. In contrast, the four pumps of Bruticus Maximus move 100,000 gallons of water per minute. The artificial waves are big enough that surfers must begin with certification runs that use only two of the pumps.

Most virtual environments are less immersive and involve watching computer screens, often in a gaming environment. Pesce (2000) traces the history of video games in his book *The Playful World: How Technology Is Transforming Our Imagination*. It began with Nolan Bushnell, a game called Pong, and Atari, the company that Bushnell founded to build his video games. Computer graphics had entered the mainstream of popular culture, which eventually became a problem for Atari as other companies fought for their share of the market. At the beginning of the 21st century, the video game industry generated over $7 billion in revenues in the United States.

One of the more successful games—SimCity—was created by Will Wright, who in the mid-1980s decided that building cities could be as exciting (and more educational) than fighting simulated battles. He was correct, and in 1997 sold SimCity to gaming giant Electronic Arts. By early 2006, Electronic Arts had sold 58 million copies of The Sims, making it the best selling video game of all time. Players can choose characters, then build houses and care for them. Today, The Sims is one of the few major video games that have more girls as registered users than boys. Electronic Arts recently donated the use of its Sims characters in a programming course at Carnegie Mellon University in a joint effort to attract more young women to computer programming. Students in the Carnegie Mellon course will write computer code to move the animated Sims characters around the computer screen.

Psychologists have also created virtual environments to help treat phobias. Several years ago, I tried an interesting demonstration at the American Psychological Association convention in Chicago that simulated walking across suspension bridges. A clinical psychologist might use a desensitizing procedure in real environments that begins with a short, sturdy bridge and gradually works up to a long, scary bridge. San Diego has enough canyons that a psychologist could likely find such a variety of suspension bridges but she and her patient would have to travel to various parts of the city. Some cities would not have any suspension bridges. The simulation utilized earphones, a movable platform, and goggles showing a movie of a suspension bridge over a deep canyon. By turning a dial, I was able to gradually increase the swaying of the bridge on the screen, the howling of the wind in my earphones, and the rocking of the platform beneath my feet.

EXPLORING ENVIRONMENTS

An important theoretical question is whether the information-processing models, such as the one depicted in Figure 4.1, are adequate for describing how people explore these and other environments. Information-processing models were encouraged by Ulric Neisser's 1967 book *Cognitive Psychology,* but Neisser subsequently became concerned about the direction of cognitive psychology. One of his concerns expressed in his 1976 book *Cognition and Reality* was that the experimental paradigms invented by cognitive psychologists were not leading the field in a productive direction. He believed that some of the paradigms, such as those that presented visual patterns too rapidly to be fully recognized, were too artificial and unreal. People usually have sufficient time to recognize objects when they explore environments.

A second concern was whether cognitive psychology placed too much emphasis on constructing stimuli and not enough emphasis on extracting information from stimuli. Neisser's (1967, p. 4) own definition of cognitive psychology as referring "to all processes by which the sensory input is transformed, reduced, elaborated, stored, recovered, and used" contributed to the emphasis on changing the sensory input, rather than simply perceiving it.

There is a potential problem with this construction view of perception that Neisser raises in *Cognition and Reality*:

Virtual Reality

> If percepts are constructed, why are they usually accurate? Surely perceiving is not just a lucky way of having mental images! The answer must lie in the kind and quality of optical information available to the perceiver. The information must be structured enough in most cases to ensure that the constructed percept is true to the real object. But if this admitted, the notion of "construction" seems almost superfluous. One is tempted to dispense with it altogether, as J. J. Gibson has done. (Neisser, 1976, p. 18)

This quote reflects the influence of Neisser's colleagues at Cornell University—James and Eleanor Gibson—to whom *Cognition and Reality* is dedicated. J. J. Gibson's theory of perception is that the organism is tuned to sensory properties of its environment that are objectively present and accurately perceived. Gibson (1966) believed that investigators of perception should develop richer descriptions of how organisms perceive (pick up) this stimulus information rather than propose hypotheses about mental constructions.

Although Neisser was greatly influenced by the Gibsons' arguments, he was not totally convinced. He believed that organized knowledge structures called schemata are needed to direct the pickup of perceptual information. As he explains in the preface to *Cognition and Reality*:

> To their dismay, I have found it necessary to suppose that the perceiver has certain cognitive structures, called *schemata* that function to pick up the information that the environment offers. This notion is central in my attempt to reconcile the concepts of information processing and information pickup, both of which capture too much of the truth to be ignored. In addition, it offers a connecting link between perception and the higher mental processes. (Neisser, 1976, p. xii)

Neisser combined the information-processing and Gibsonian approaches by creating the perceptual cycle shown in Figure 12.1. The available sensory information is contained in the object but the schema controls the activity of looking. It contains plans for finding out about objects and events by directing exploratory movements of the head and eyes. The information that is perceived modifies the schema, which directs further exploration.

ECOLOGICAL PSYCHOLOGY AND MULTIMEDIA

Neisser's perceptual cycle is an important building block in the construction of a theoretical foundation for a field of psychology known as *ecological psychology*. Michael Young (2004) at the University of Connecticut describes this field and its implications for instructional design in his chapter in the *Handbook of Research for Educational Communications and Technology*. Young contrasts cognitive and ecological psychology by referring to two metaphors. The favorite metaphor of cognitive psychology is that people are like computers taking in, storing, and retrieving information from short- and long-term memory.

Young (2004) selects the thermostat as a metaphor for ecological psychology. The thermostat is a control device with a goal (maintain the room at a set

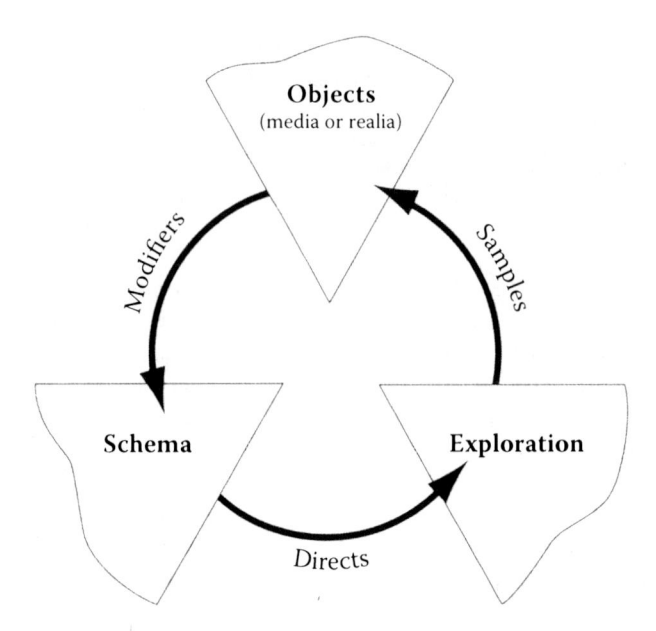

FIGURE 12.1 Neisser's perceptual cycle. (From *Cognition and Reality* (p. 21), by U. Neisser, 1976, San Francisco: W. H. Freeman and Company. Copyright 1976 by W. H. Freeman and Company. Reprinted with permission.)

temperature). It interacts continuously with the environment to measure the room's temperature and reacts when the temperature departs too far from the set temperature. The most critical attributes of this metaphor are that the interaction is dynamic and continuous, not static or linear as in many information-processing models, and the perceiving-acting cycle unfolds as a feedback loop with control parameters and action parameters. These attributes are captured in Neisser's (1976) perceptual cycle.

Allen, Otto, and Hoffman (2004) also contributed a chapter to the same handbook in which they apply the principles of ecological psychology to educational technology. They argue that the communications metaphor of symbols, messages, and discourse fails to include some important aspects of how people interact with environments. These interactions often require locating, tracking, identifying, grasping, moving, and modifying objects.

The authors refer to our interactions with desktop computers as an example of how symbols on the screen differ from the symbols that are used in communication. Communication symbols such as words refer to objects and events but are not objects for initiating events. In contrast, screen icons allow us to manipulate objects by pointing, clicking, and dragging. We can open and close documents, place them in files, move the files around the screen, and throw unwanted documents into trashcans on our virtual desktop.

We can also do much more with our desktop computers as illustrated by Bob Hoffman's CD-ROM, *The Mystery of the Mission Museum* (2000). The CD uses virtual reality to make one of the 18th century California missions (La Prisima at Lompoc) more accessible to students. The virtual mission consists of 176 photographically generated, 360° panoramas in which viewers can move through the mission rooms and gardens (Figure 12.2). They can interact with many of the objects in the rooms by using the computer mouse in approximately the same way that people would interact with real objects. For example, students can operate a spinning wheel by clicking on the wheel and moving it in a circle. They can also lead the olive-mill and wheat-mill donkeys, point and shoot the mission's cannon, and pull the mission bell rope.

Neisser's perceptual cycle and other work within the ecological psychology framework have made important contributions for specifying the role of vision and action in obtaining information from real and virtual environments. It is helpful to contrast the contributions of Neisser's *Cognition and Reality* (1976) with Piaget's *Recherches sur l'abstraction réfléchissante* (1977), which was published (in French) one year later. Neisser emphasized the relationship between perception and action, whereas Piaget emphasized the relation between action and reflection. Both are important in understanding visual thinking. Understanding perception is necessary for understanding the visual part of visual thinking, and understanding reflection is necessary for understanding the thinking part of visual thinking. Both perception and thinking play key roles in making decisions in complex, changing environments, as discussed in the next sections.

SITUATION AWARENESS

A common theme that runs through many virtual training or testing environments is to measure or increase people's awareness of their situation. Measuring situation awareness was the goal of a project in which I participated at the Naval Personnel Research and Development Center. It was located on Point Loma—the place where photographers take those picture postcards of sailboats in the foreground, the San Diego Harbor in the middle, and the city skyline in the background. The summer was 1993—several years before the center fell victim to base closures.

My assignment as a visiting faculty member was to assist on a study of electronic warfare technicians under the direction of Josephine Randel and Larry Pugh (Randel, Pugh, & Reed, 1996). Electronic warfare technicians use high-tech computer equipment to detect and identify signals from numerous, potentially hostile, radar systems. The technicians try to determine how the radar is being used by listening to auditory signals and observing a computer-generated diagram of the position of all the radar emitters in the area. The movie *The Hunt for Red October* is an example, although instead of tracking a single submarine there can be many ships and planes in a battle area.

The workload of the technician in a battle situation can therefore become very demanding because of the large amount, fragmented nature, and unreliability of the collected information. Decisions on the sources of the radar emitters, their significance, and possible countermeasures must be made very quickly. These

FIGURE 12.2 Screens from *The Mystery of the Mission Museum*. (From "Media as Lived Environments: The Ecological Psychology of Educational Technology," by B. S. Allen, R. G. Otto, and B. Hoffman. In *Handbook of Research on Educational Communications and Technology* (2nd ed., pp. 215–241), by D. H. Jonassen (Ed.), 2004, Mahwah, NJ: Lawrence Erlbaum Associates. Copyright 2004 by Lawrence Erlbaum Associates. Reprinted with permission.)

decisions depend on situation awareness of the movement of planes and ships in the area, both hostile and friendly.

The theoretical framework for our study was the recognition-primed decision model proposed by Klein (1993). Klein initially formulated his model after interviewing ground commanders about how they made decisions when fighting forest fires. Rather than evaluate many alternative courses of action, they reported that they used their prior experience to immediately generate and then modify plans in reaction to the changing situation caused by changes in wind, humidity, terrain, and vegetation. Klein's model is called a *recognition-primed* model because of the emphasis it places on recognition of events in assessing the situation. Once the problem is recognized, experienced decision makers can usually quickly identify an acceptable course of action.

The key to making good decisions, therefore, is to understand the situation as it constantly changes. Randel et al. (1996) studied electronic warfare technicians' ability to do this by providing visual and auditory signals in a 35-minute training scenario involving action in the North Pacific. The training scenario involved both hostile and friendly radar emitters. We were particularly interested in how expertise would influence situation awareness. One of the measures involved stopping the simulation, giving technicians a drawing of the blank computer screen and asking them to draw from memory the location of the symbols representing the different radar emitters.

Randel et al. (1996) found two major results. The first was that, as predicted by Klein's (1993) model, experts were much better than intermediates were, and intermediates were much better than novices were, at reproducing the screen. The second was that technicians at all levels of expertise were better at remembering the location of hostile emitters rather than friendly emitters, demonstrating the differential importance of the two kinds of signals. The experts correctly drew 95% of the locations of the hostile emitters, compared to 78% for the intermediates, and 51% for the novices. The experts correctly drew 82% of the locations of the friendly emitters, compared to 53% for the intermediates, and 21% for the novices.

In addition to the lower situation awareness of the novices, the novices placed less emphasis on understanding the situation before making a decision. Detailed interviews of the electronic warfare technicians at the end of the exercise found that eight of the nine experts, all thirteen intermediates, and only two of the six novices emphasized the importance of situation assessment before selecting a course of action.

VIRTUAL MILITARY TRAINING

Situation assessment can be increased with training in virtual reality environments when training in real environments would be costly or dangerous. The University of Southern California's Institute for Creative Technologies, with help from Hollywood, is currently using a 5-year $100 million grant from the Pentagon to create a kind of theme-park version of a war zone. The training is lead by Sgt.

John Blackwell, a life-sized three-dimensional simulation of a person whose mission is to train real soldiers. The soldiers are placed in a virtual world that consists of confusing, chaotic situations to test and enhance their ability to quickly make sense of it.

The training scenario the group is currently using depicts a peace-keeping mission in a small town in Bosnia. It begins with the trainee, depicted as a lieutenant, receiving orders to assist in controlling a civil disturbance. On the way to the disturbance, he discovers that a small boy has been injured during a collision with one of his platoon's Humvees (Figure 12.3). Should the lieutenant render aid or continue with his mission (a dilemma that would later play a central role in the movie *In the Valley of Elah*)?

The virtual system allows the trainee to interact through speech with three virtual humans—the sergeant, a medic, and the mother of the small boy. The interactions end with one of four possible consequences that depend on the trainee's actions. The capabilities of the virtual humans are limited by their ability to understand both speech and tasks. The sergeant currently has the ability to understand a few hundred words with a grammar allowing for recognition of 16,000 distinct utterances. He also has knowledge of about 40 different tasks.

As expected, the ability to make good decisions depends on the trainee's military knowledge, but the training scenarios provide an environment in which to acquire this knowledge. As work continues, the virtual environments take on new capabilities. One of the most recent is the training of negotiation skills. Another is the use of virtual reality to treat veterans suffering from posttraumatic stress disorder. These activities are still under development but the initial findings are promising (Rizzo et al., 2009).

FIGURE 12.3 A virtual environment that allows for interaction with the platoon sergeant, the mother of an injured boy, and a medic. From "Toward Virtual Humans," by W. Swartout, R. H. Gratch, R. Hill, E. Hovy, S. Marsella, J. Rickel, et al., 2006, *AI Magazine*, 27, 96–108. Copyright 2006 by American Association for Artificial Intelligence. Reprinted with permission.

VIRTUAL BIKING

Fortunately, however, we already have results from studies that have used virtual environments to study perception and action in everyday activities such as riding a bicycle and driving a car. These activities also involve risk. Approximately 600,000 bicycle-related injuries in the United States are treated in emergency rooms each year, with 5- to 15-year-old children representing a particularly high percentage of that group.

At the University of Iowa, Plumert, Kearney, and Cremer (2007) have created a virtual environment to study how immature perceptual-motor functioning can put children at risk when crossing roads. Children ride a bicycle mounted on a stationary frame that can simulate mass and inertia, terrain slope, ground friction, and wind resistance (Figure 12.4). The bicycle is located in the middle of three 10-foot by 8-foot screens that provide high-resolution visual information. In one study, 10- and 12-year-olds and adults rode the bicycle on a residential street with six intersections. Their goal was to cross the intersections without being "hit" by a car. This required judging whether the time gap between two cars would provide sufficient time to ride the bike across the intersection.

Children and adults chose the same size gaps between cars but children had less time to clear the path of the approaching car. They delayed in beginning

FIGURE 12.4 A virtual environment to study bicyclists' decisions to cross a busy intersection. (Based on "Children's road crossing: A window into perceptual-motor development." by J. M. Plumert, J. K. Kearney, and J. F. Cremer, 2007, *Current Directions in Psychological Science, 16*, 255–263. Copyright by Blackwell. Reprinted with permission.)

to peddle relative to adults and often overestimated their physical abilities. The margin of error was very small for 10-year-olds, particularly for 10-year-old girls. Children selected smaller gaps when they had to wait longer to cross, resulting in less than a second to spare when avoiding an approaching car.

VIRTUAL DRIVING

We hope that approaching cars would slow down in real environments, but inexperienced drivers are not always careful observers. Per 100 million vehicle miles, a 16-year-old is almost eight times as likely to get into a fatal crash as a 45- to 64-year-old, and an 18-year-old is four times as likely. There is clearly a need for training programs that could provide needed experience.

Pollatsek, Fisher, and Pradhan (2006) designed such a training program that requires less than an hour to complete on a personal computer. The software requires dragging red circles to areas of the roadway that should be continually monitored and dragging yellow circles to areas that could contain relevant hidden information such as pedestrians emerging behind hedges. The training involved both coaching and review tests.

The investigators then used the University of Massachusetts driving simulator to evaluate the success of the training. The simulator uses a 1995 Saturn sedan and a virtual world projected onto three screens surrounding the car, similar to the one depicted in Figure 12.4. Participants control the vehicle in the same way that they would control a normal vehicle. A head-mounted eye tracker records where the drivers look as they navigate through the virtual world.

Pollatsek et al. (2006) studied a group of novice (16- and 17-year-old) drivers on the simulator to measure whether they would attend to critical areas. The trained drivers fixated the appropriate region 58% of the time whereas the untrained drivers fixated it only 35% of the time. A follow-up experiment revealed that there was no decrement in the training when the test on the driving simulator was given 3 to 5 days after training. A field study in the Amherst, Massachusetts, environment also showed that training transferred to real environments. The eye tracker revealed that trained drivers looked at critical areas 64% of the time, compared to 37% for untrained drivers.

The creation of virtual environments to study and improve perception, action, learning, and decision making is beginning to produce results. I hope to soon see more widespread use of virtual reality, particularly to help people cope with risky situations. Training drivers, for example, is important because they typically underestimate how increasing speed increases risk (Svenson, 2009).

SUMMARY

Virtual reality creates physical simulations of real world activities that provide entertainment, training, or both. Examples include constructing simulated cities, exploring virtual museums, taking simulated driving tests, and training soldiers in simulated environments. Theories of these activities, such as Neisser's

(1976) perceptual cycle, describe a more cyclical interplay between perception and action than is emphasized in information-processing models that are more traditional. One example is Klein's (1993) recognition-primed model in which situation awareness is the key component of making good decisions in risky environments. Situation awareness typically depends on visual perception—whether it is driving down a road, providing assistance following an accident, updating the location of signals on a radar screen, or monitoring the progression of a forest fire. It is also necessary to select a good course of action, but good decisions begin by understanding the situation. Training and experience increase situation awareness, as illustrated by the training of novice drivers and the evaluation of electronic warfare technicians.

13 Science Instructional Software

A book on visual thinking would be incomplete without some discussion of how to promote and encourage it in schools. Previous chapters have provided examples of how the manipulation of physical objects can help young children get started in reading and mathematics. Examples include Glenberg's (2004, 2007) use of toy farms and toy zoos to assist children in reading and solving simple mathematics problems. However, computer technology provides greater power by providing microworlds that students can explore. *Microworlds* are small virtual (computer-based) worlds with which people can interact in order to learn. The previous chapter on virtual reality provided examples, such as The Sims.

Lloyd Rieber at the University of Georgia begins his chapter on microworlds with a description of how they differ from the initial developments of instructional software:

> The introduction and spread of computer technology in schools since about 1980 have led to a vast assortment of educational software. Most of this software is instructional in nature, based on the paradigm of "explain, practice, and test." However, another, much smaller collection of software, known as *microworlds*, is based on very different principles, those of invention, play, and discovery. Instead of seeking to give students knowledge passed down from one generation to the next as efficiently as possible, the aim is to give students the resources to build and refine their own knowledge in personal and meaningful ways. (Rieber, 2003, p. 583)

Rieber (2003) reviews the development of microworlds in a historical context, beginning in the year 1980. This year is important for two reasons. First, the Apple II had just been introduced, making it possible for educators to use computers in a typical classroom. Second, the publication of Seymour Papert's book *Mindstorms* (1980) pointed to a new direction for educational software. Papert developed a programming language called Logo that enabled young children to move a turtle around a computer screen by using commands such as FORWARD, BACK, LEFT, and RIGHT. Users could now construct geometric figures from the trail left by the turtle. The word "construct" is important because Papert argued that microworlds should allow students to participate in active, exploratory learning in an environment that is sufficiently rich for interesting discoveries. Trained as a mathematician but influenced by Piaget's constructivist philosophy, Papert created a microworld that offered a constructivist approach to learning mathematics.

This chapter on scientific software provides examples of computer programs that offer instruction in physics, ecology, chemistry, and experimental design. The programs differ in the extent to which they allow students to construct knowledge by exploring microworlds, but all encourage visual thinking. They illustrate a variety of scientific content and instructional methods for teaching that content.

PHYSICS

A major early contribution to the study of microworlds was Barbara White's dissertation that she submitted to the Massachusetts Institute of Technology in 1981. The microworld (White, 1984) was initially created by Andy diSessa and inspired by Papert's use of technology to support a constructivist approach to learning. The goal of White's research was to create a sequence of games that would enable students to learn Newton's first two laws of motion in an idealized (frictionless) environment. As you may recall, Newton proposed that

1. In the absence of forces, objects at rest stay at rest and objects in motion maintain uniform motion in a straight line.
2. A force applied to an object accelerates the motion of the object in proportion to the size of the force and mass of the object (F = MA).

Students were told to imagine a spaceship traveling through space so there is no friction. An engine that gives a sudden burst of force and then shuts itself off drives the spaceship. Two important considerations in designing the games were to expose students' misconceptions and provide feedback that could help correct their misconceptions. Figure 13.1 illustrates a common misconception. The arrows (vectors) show the direction of movement and their lengths show the speed of movement. Many students initially believe that the spaceship will head in the direction of the impulse so they apply an upward impulse to make the spaceship

(a) Expected Result   (b) Actual Result

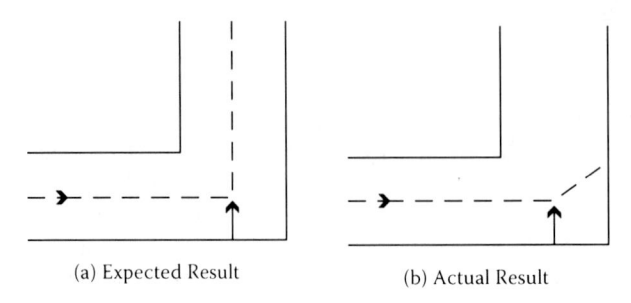

FIGURE 13.1 A common expected result (a) and the actual result (b) when a vertical impulse is applied after a horizontal impulse. (From "Designing Computer Games to Help Physics Students Understand Newton's Laws of Motion," by B. Y. White, 1984, *Cognition and Instruction, 1*, 69–108. Copyright 1984 by Lawrence Erlbaum Associates. Reprinted with permission.)

Science Instructional Software 149

turn the corner (Figure 13.1a). The spaceship instead turns at a 45° angle and crashes into the wall (Figure 13.1b).

White discovered that after less than one hour of playing the games, students in a high school physics class improved their understanding of Newton's laws. They gave more correct answers to questions about stopping the spaceship, turning right corners, and moving in a circular path. The findings also revealed limitations of the initial design. Many students focused only on the direction of movement and did not notice changes in speed because they did not expect to see changes in speed. Improving the design by having the spaceship leave a physical trace of its path (as in Papert's turtles) and providing a digital display of speed made Newton's laws more obvious.

White continued to improve her design after obtaining a faculty position at the School of Education, University of California, Berkeley. Her improved design, labeled ThinkerTools, is illustrated in Figure 13.2 (White, 1993). The goal of this particular task is to make the object navigate the track and stop on the target. The speed and direction of the object's motion are shown by the simulation and by the

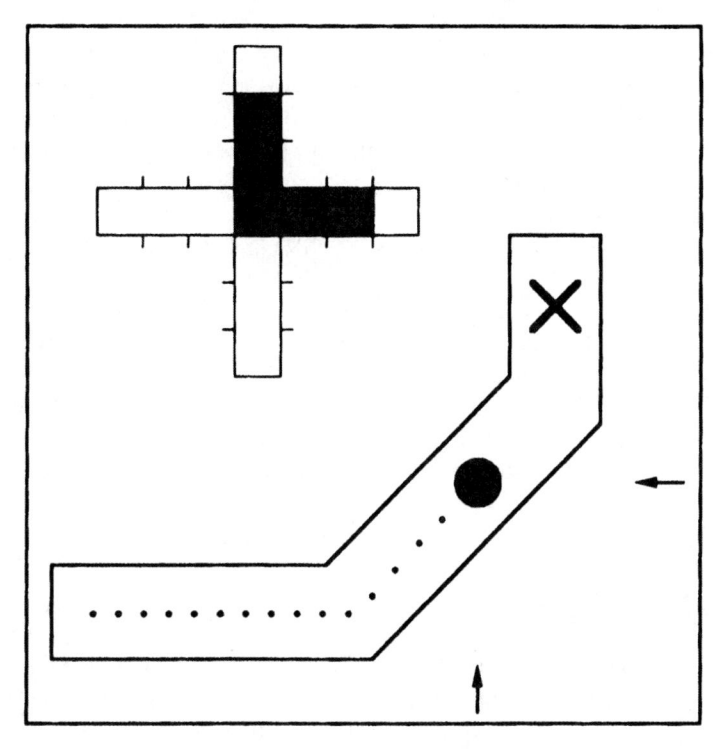

FIGURE 13.2 A screen from the ThinkerTools software that shows a moving circle, its wake, and a datacross depicting the speed and direction of the two impulses. (From "ThinkerTools: Causal Models, Conceptual Change, and Science Education," by B. Y. White, 1993, *Cognition and Instruction, 10,* 1–100. Copyright 1993 by Lawrence Erlbaum Associates. Reprinted with permission.)

physical trace of its path that leaves small dots on the screen at fixed time intervals. Figure 13.2 reveals that the object began moving horizontally but turned the 45° corner after the application of a vertical impulse that had the same speed as the horizontal impulse. The datacross in the upper left of the screen shows the direction and speed of these two impulses. The trail of dots reveals the object's change in direction and speed after the second impulse. The dots are further apart after turning the corner, indicating an increase in speed.

White (1993) evaluated the effectiveness of the ThinkerTools software by designing increasingly challenging tasks for a group of sixth-grade students in a middle-class Boston suburb. At the end of the 2-month curriculum, students took a transfer test that measured their understanding of Newtonian mechanics in real-world problem contexts. The test included predictive questions that frequently generated wrong answers from high school and college students. The sixth graders who interacted with ThinkerTools did significantly better in answering these questions than a control group of high school students who been taught about force and motion using traditional textbook methods.

A more recent application of the ThinkerTools software has been in the context of the Model-Enhanced ThinkerTools curriculum, which enables middle school students to create and test scientific models. Schwarz and White (2005) evaluated the curriculum in four seventh-grade classrooms for approximately 45 minutes a day for 10.5 weeks. The students collected data from real world-experiments during the initial phase, such as measuring the speed of a plastic puck over two different surfaces that varied in the amount of friction. They then used the instructional software to formulate different models to see which model best fit their data (such as friction causes objects in motion to slow down and eventually stop).

After using the curriculum, most students had learned that scientific and computer models are useful in a variety of ways including visualization, testing theories, making predictions, and helping people understand science. Most students also understood that there could be multiple models of the same phenomena and models are estimates of the physical world. On the other hand, the results also suggested that many students thought all models were of equal value. Schwarz and White (2005) speculated that the particular culture of the school, which emphasizes that everyone's ideas must be respected, may have contributed to the lack of improvement on questions about the comparative value of alternative models.

ECOLOGICAL SYSTEMS

In 1992, another graduate student at MIT completed a dissertation with a different approach to computer modeling. Its author, Mitchell Resnick, believed that emphasis should also be placed on developing microworlds that are *not* guided by a central plan. His alternative to centralized thinking is ecological thinking in which plans gradually develop as responses to the environment. We have already seen examples of ecological thinking in Young's (2004) analogy of a thermostat, Neisser's (1967) perceptual cycle, and Klein's (1993) recognition-primed model of decision-making.

Resnick (2003) later argued that many organisms, such as the "walking tree" in Costa Rica, exhibit ecological planning. The walking tree sits atop a bundle of roots that raise it about 1 meter above the ground. It can gradually shift its position by following three rules:

1. Test the soil randomly by sending roots out in all directions.
2. Determine which roots find the best soil.
3. Use this information to determine in which direction to move.

This strategy shares two common characteristics of ecological thinking—respond to local conditions and adapt to changing conditions.

Responding to the environment may also include reacting to other objects such as vehicles on a freeway. To model these more complex kinds of interactions, Resnick (2003) expanded Papert's Logo language so it could simulate the movement of many objects. In one application of the enhanced language (StarLogo), two high school students were able to space cars evenly along a highway by calculating the distances between cars. One of Resnick's graduate students then designed an alternative plan that is more characteristic of ecological thinking. Each car repeatedly responded to the following rules:

1. Calculate the distance to the car in front of me.
2. Calculate the distance to the car behind me.
3. Take a step toward whichever one is further away.

A classic case of ecological modeling involves fluctuations in the population sizes of predators and prey. One mathematical model that has had a fair amount of success predicts that the sizes of both populations fluctuate like two displaced sine waves. Increasing the number of prey increases the number of predators, which decreases the number of prey, which decreases the number of predators, which increases the number of prey, and so on. These population dynamics are difficult to model unless one is familiar with differential equations.

However, Uri Wilensky at Northwestern University created a more sophisticated successor to StarLogo (called NetLogo) that uses rules that an *individual* predator or prey must follow to generate these changes in the two populations. Wilensky and Reisman (2006) describe a case study of a student named Talia whose interaction with NetLogo is typical of high school students. Talia initially created four rules to describe the behavior of a wolf and two rules to describe the behavior of a sheep.

Rule set for a wolf:

1. Move randomly to an adjacent patch and decrease energy by E_1.
2. If on the same patch with one or more sheep, eat a sheep and increase energy by E_2.
3. If energy < 0, then die.
4. With probability R_1, reproduce.

Rule set for a sheep:

1. Move randomly to an adjacent patch.
2. With a probability of R_2, reproduce.

Talia next experimented with different values of the four parameters (E_1, E_2, R_1, R_2) and discovered that either of two outcomes would always occur. One outcome (shown in Figure 13.3a) increased the sheep population, which increased the wolf population, which caused the extinction of the sheep, which caused the extinction of the wolves. The second outcome (shown in Figure 13.3b) resulted in the sheep barely escaping extinction before the extinction of the wolf population,

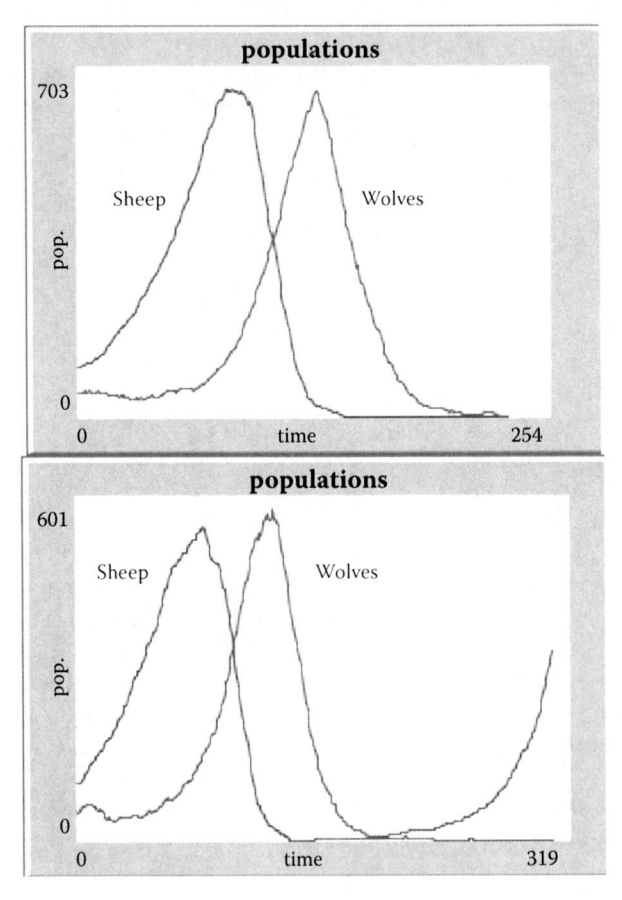

FIGURE 13.3 Two different outcomes from varying parameter values in Talia's initial rule set. (From "Thinking like a wolf, a sheep, or a firefly: Learning biology through constructing and testing computational theories—an embodied modeling approach," by U. Wilensky and K. Reisman, 2006, *Cognition and Instruction 24*, 171–209. Copyright by Lawrence Erlbaum Associates. Reprinted with permission.)

which created unlimited growth of the sheep. Talia was eventually able to design a set of rules that would produce realistic changes in the two populations. The new set of rules for a sheep corresponded to the four rules for a wolf. The abundance of grass curtailed the growth of the sheep population in the same way that the abundance of sheep curtailed the growth of the wolf population. Talia's example illustrates how, by using NetLogo, precalculus students can understand and model population dynamics without the necessity of knowing differential equations.

TRANSFER OF PRINCIPLES

An advantage of modeling ecological systems is that common principles can apply across different situations. Spacing cars along a freeway, for example, has some similarities with animals distributing themselves over a territory so they will not compete for food resources. Two cognitive psychologists, Robert Goldstone and Ji Son (2005), became interested in the learning and transfer of such principles across different tasks. Their first task, shown in Figure 13.4, allows users to control how ants move toward food sources based on three rules:

1. One at a time, a piece of food is randomly selected from all the food present.
2. The ant that is closest to the selected food moves toward the food with a speed that you specify.
3. All of the other ants move toward the food with another speed.

Undergraduates at Indiana University explored the simulation by drawing different food patches, changing the number and position of the ants, and controlling their speed. Their goal, however, remained the same—try to move each ant into its own food patch. After students explored how to accomplish this goal, they were given a new task that required learning patterns. Pattern learning was governed by the same rules as the food task, and the investigators were interested in determining how easily students could transfer the principles from the first to the second task.

Transfer is challenging because the two tasks look very different. Therefore, it is not obvious how to generalize the principles that govern the behavior of ants to learning the patterns. Previous research suggested to Goldstone and Son (2005) that transfer would be easier if the initial task were presented more abstractly by using small black dots instead of ants and green patches instead of food. The green patch in this abstract condition was referred to as a green dot, the closest black dot was referred to as the winner, and the other black dots were referred to as the losers.

The concrete and abstract versions of the initial task allowed for four variations of training. One variation used only the concrete task, a second used only the abstract task, a third switched from the concrete to the abstract task, and a fourth switched from the abstract to the concrete task. The results demonstrated that the students were more successful in learning the principles governing the

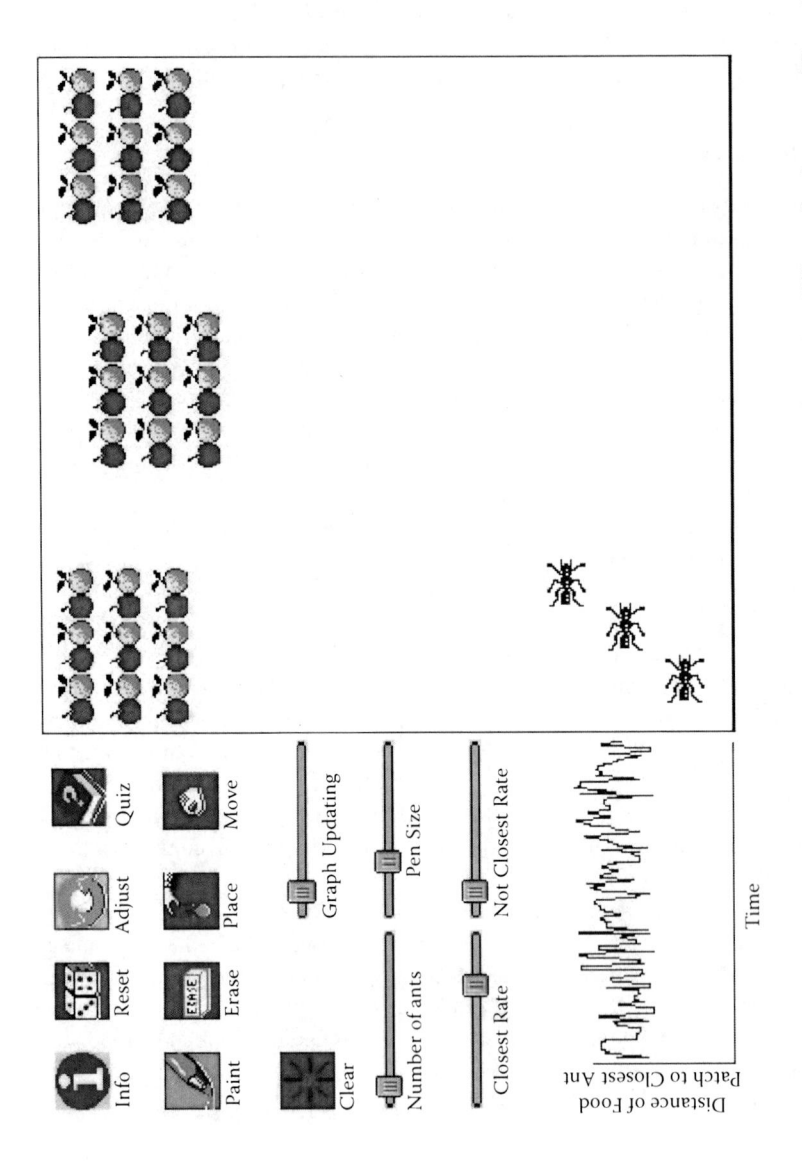

FIGURE 13.4 A screen from the ants and food simulation in which the goal is to move the ants to different food sources. (From "The Transfer of Scientific Principles Using Concrete and Idealized Simulations," by R. L. Goldstone and J. Y. Son, 2005, *Journal of the Learning Sciences, 14,* 69–110. Copyright 2005 by Taylor & Francis Group. Reprinted with permission.)

Science Instructional Software

transfer task if they began training with the concrete (ants) task and then switched to the more abstract task. Goldstone and Son (2005) conclude:

> Consistent with the advantages of concrete representations described in Table 1, we believe, along with many others, that computer simulations are effective pedagogical devices precisely because of their concreteness and perceptual grounding. However, we are also interested in students applying what they have learned to domains that are superficially unrelated to the simulation's domain. Our results give us optimism that these motivations are not necessarily mutually exclusive and that both perceptually grounded and abstract understandings can be simultaneously achieved. (p. 101)

Table 13.1 is a replica of Table 1 referred to in the previous quotation. It compares the advantages of concrete and idealized (abstract) representations.

There is a parallel between the abstract shapes that Goldstone and Son (2005) created and the circle that White (1993) created in her ThinkerTools curriculum. White intentionally used the circle in Figure 13.2 to depict that Newton's laws of motion apply to all objects in a frictionless environment. The circle therefore abstractly represents a large variety of concrete objects.

Table 13.1 Advantages of Concrete and Idealized Representations

Advantages of Concreteness	Advantages of Idealization
Concrete information is easier to remember than abstract information.	Idealizations are potentially more transferable to dissimilar domains because knowledge is not as tied to a specific domain.
It is often easier to reason with concrete representations using mental models than abstract symbols.	The critical essence of a phenomenon is highlighted because distracting details are eliminated.
Visual processes used for concrete objects can be co-opted for abstract reasoning.	There may be an active competition between treating an entity as a symbol versus an object, and idealization makes symbolic interpretations more likely.
Concrete details are not always "superficial," but rather provide critical information about likely behavior and relevant principles.	Cognitive processing of less important but complex concrete elements is conserved.
Concrete materials are often more engaging and entertaining and less intimidating.	Idealizations facilitate interpretations of a situation in terms of abstract relations rather than specific attributes.
Concretely grounded representations are more obviously connected to real-world situations.	

From: "The Transfer of Scientific Principles Using Concrete and Idealized Simulations," by R. L. Goldstone and J. Y. Son, 2005, *Journal of the Learning Sciences, 14,* 69–110. Copyright 2005 by Taylor & Francis Group. Reprinted with permission.

CHEMISTRY

Chemistry is another science in which visual thinking is important. A review of research on visuospatial thinking in learning chemistry by Hsin-Kai Wu at the National Taiwan Normal University and Priti Shah at the University of Michigan supports their claim that chemistry is a visual science (Wu & Shah, 2004). The research reveals that students with lower visuospatial abilities

1. are unable to perform as well as students with higher visuospatial abilities on solving both spatial and nonspatial chemistry problems and
2. have difficulty transforming the information in questions into a visual representation, such as drawing preliminary diagrams.

In addition, many students are unable to

1. form three-dimensional mental images by visualizing two-dimensional structures;
2. make translations between a chemical formula, electron configuration, and a ball-and-stick model; and
3. visualize the interactive and dynamic nature of a chemical process by viewing symbols and equations.

Wu's prior research with Krajcak and Soloway (2001) at the University of Michigan investigated the use of a visualization tool, eChem, that enabled students to build molecular models and view multiple representations of these models. eChem guides students in building molecules, visualizing multiple three-dimensional views, and relating the molecules to concepts in chemistry. Figure 13.5 shows the graphic interface of the Visualize page for the ethanol molecule. The visualize feature allows students to compare various representations such as ball-and-stick, filled-space, and wire-frame models.

Wu's study examined how different features of eChem could help students develop their ability to visualize and make translations between chemical representations (Wu et al., 2001). Seventy-one eleventh-grade students at a small, public high school in the Midwest that was working with educational researchers to develop a 3-year, inquiry-based science curriculum were involved in the study. Students used the software during a 6-week unit on toxins to construct three-dimensional models, post their models on their Web pages, and justify their arguments about polarity, solubility, and toxicity of the different chemical structures.

The investigators collected multiple sources of data during the 6 weeks including video recordings of classroom interactions, field notes of classroom observations, and interview transcripts. In addition, a comparison of a pretest and posttest revealed that students' understanding of chemical representations improved substantially (the average posttest score was over 2.5 standard deviations above the average pretest score). The investigators concluded that their visualization tool

Science Instructional Software 157

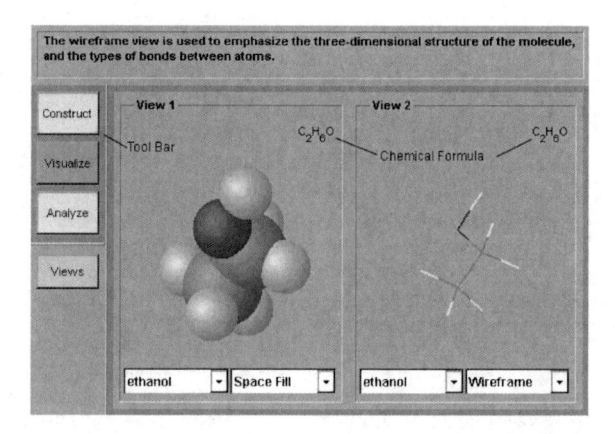

FIGURE 13.5 A screen from eChem depicting a space-filled and a wire-frame view of the ethanol molecule. (From "Promoting Understanding of Chemical Representations: Students' Use of a Visualization Tool in the Classroom," by H.-K. Wu, J. Krajcik, and E. Soloway, 2001, *Journal of Research in Science Teaching, 38*, 821–842. Copyright 2001 by *Journal of Research in Science Teaching*. Reprinted with permission.)

encourages students to discuss how the different visual representations of molecules relate to the underlying concepts of chemical bonding.

Other software has focused on the dynamic processes of chemical reactions (Kozma & Russell, 2005). Robert Kozma at the Stanford Research Institute's Center for Technology in Learning compared how novices (chemistry students) and experts (chemists) established connections across four different representations: chemical equations, graphs, molecular-level animations, and video of laboratory experiments (Kozma et al., 1996). The chemists made connections across these different representations based on principles such as "collision theory" and the "gas law." However, the students found it difficult to make connections because they focused more on the physical characteristics of the displays such as "molecules moving about" and "graphs of concentrations."

Kozma and his colleagues (1996) therefore developed the instructional software shown in Figure 13.6 to help students make connections across the four representations. The software, MultiMedia and Mental Models or 4M:Chem, creates multiple, linked representations to illustrate chemical equilibrium. Students can select a chemical system such as the one in Figure 13.6 that illustrates a chemical reaction involving two gasses, represented symbolically as N_2O_4 (dinitrogen tetroxide) and NO_2 (nitrogen dioxide). It is the brownish color of nitrogen dioxide that can be seen as the principle component of smog.

Software users can observe what happens as they heat the mixture by clicking on the letter *V* to see a video of the lab experiment, the letter *A* to see an animation of the molecules, or the letter *G* to see a graph of the reaction. They can also click on more than one letter at a time to make connections across the different representations. For example, heating the mixture would show the mixture change to a

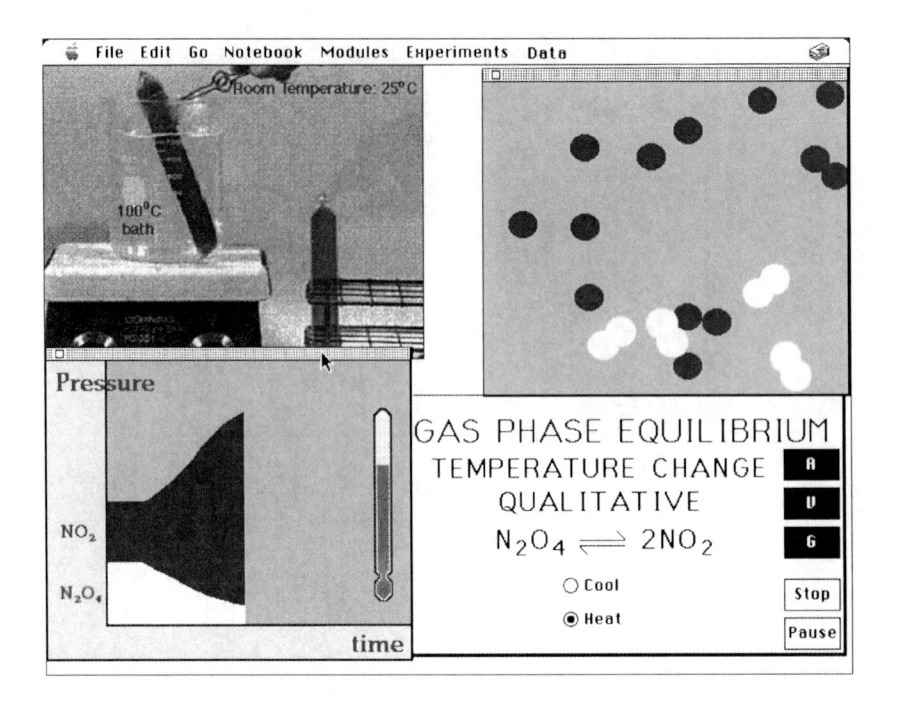

FIGURE 13.6 A screen from 4M:Chem showing multiple, linked representations of a chemical reaction. (From "The Use of Multiple, Linked Representations to Facilitate Understanding of Chemistry," by R. Kozma, J. Russell, T. Jones, and N. Marx. In *International Perspectives on the Design of Technology-Supported Learning Environments* (pp. 41–60), 1996, S. Vosniadou, E. De Corte, R. Glaser, and H. Mandl (Eds.), Mahwah, NJ: Lawrence Erlbaum Associates. Copyright 1996 by Lawrence Erlbaum Associates. Reprinted with permission.)

brownish color in the video, the conversion of white balls (N_2O_4) to brown balls (NO_2) in the animation, and the decrease of pressure for N_2O_4 and the increase of pressure for NO_2 in the graph.

Kozma and his colleagues (1996) evaluated the effectiveness of the software during two 1-hour sessions in a college chemistry course. The results were mixed. The use of 4M:Chem increased students' understandings of systems at equilibrium and the effects of temperature on those systems. However, some students still did not understand the effect of temperature, and the general understanding of the effect of pressure remained low.

The greatest challenge in having access to the variety of representations illustrated in Figure 13.6 is the visual demands of attending to multiple events, particularly because those events change over time. Ardac and Akaygun (2005) at Bogazici University in Istanbul used Sweller's (2003) cognitive load theory (discussed in Chapter 10) to design multimedia instruction showing chemical reactions using symbols, molecular animations, and video clips. They compared two groups that viewed the animations (either individually or as a class) with a class that saw 48 static transparencies selected from the animation screens. All three groups worked with handouts during five class sessions that required them to record, draw, and comment on their observations. The two groups who viewed the animations showed substantial gains from a pretest to a posttest. In contrast, the class who saw the static pictures made only modest gains.

SCIENTIFIC DESIGN

We have been examining science software across different sciences such as physics, ecology, and chemistry. However, a common theme in science is the ability to control variables in scientific experiments. Designing science experiments typically requires interacting with real objects, but can designing experiments with instructional software be as effective? Triona and Klahr (2003) studied this question at Carnegie Mellon University by instructing fourth- and fifth-grade students on how to design simple experiments by isolating and testing one variable at a time.

The experiment required students to evaluate how variables such as length, width, wire size, and weight influence the stretching of a spring. After selecting a pair of springs and weights, children clicked on GO to see a video of how far the springs stretched. The purpose of the instruction was to teach children that evaluating variables requires varying only a single variable at a time. Figure 13.7 shows a correct test for the effect of spring length because the other three variables (width, wire size, and weight) are identical.

Triona and Klahr (2003) compared a group of children who trained on the instructional software with a group of children who trained with real springs and weights. Their results showed that children who trained with the virtual materials were as capable of correctly designing experiments as were children who trained with the physical materials. Following training, both groups were asked to design experiments to evaluate the effects of four variables (steepness, length, surface, and type of ball) on the time it would take a ball to roll down a ramp. Only physical materials were used on this transfer task. Again, the group who had trained on virtual springs did as well as the group who had trained on real springs in designing experiments with real ramps, even though they had not interacted with physical materials during the training.

Another study with some of the same research team (Klahr, Triona, and Williams 2007) compared the effectiveness of constructing and evaluating toy cars in either a real or a virtual environment. Seventh- and eighth-grade students assembled and tested the cars in order to design a car that would travel the farthest. Computer-based virtual design was again equally effective and it avoided some of the problems encountered when assembling real cars. These included real cars that did not travel straight, had wheels that were too tight, and required a long corridor for testing.

The investigators concluded that their findings support the effectiveness of manipulating virtual objects. This does not imply that teachers should abandon

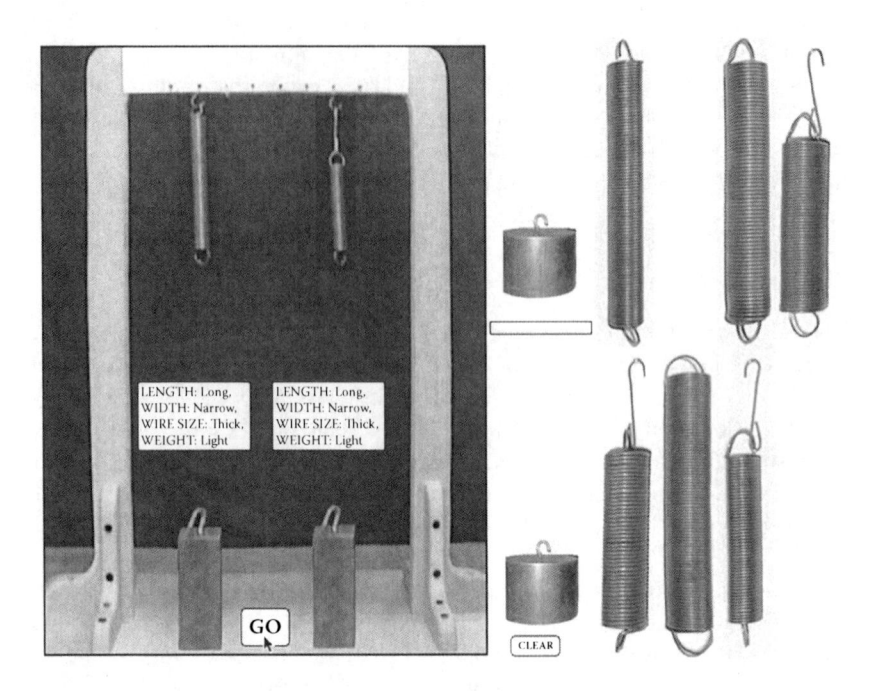

FIGURE 13.7 Virtual environment for designing experiments. (From "Point and Click or Grab and Heft: Comparing the Influence of Physical and Virtual Instructional Materials on Elementary School Students' Ability to Design Experiments," by L. M. Triona and D. Klahr, 2003, *Cognition and Instruction, 21*, 149–173. Copyright 2003 by Lawrence Erlbaum Associates. Reprinted with permission.)

hands-on science materials, but teachers should not assume that virtual materials would be less effective. There are many advantages to using computer-based laboratory materials including portability, safety, cost-efficiency, and flexible, rapid, and dynamic data displays.

SUMMARY

Animation has played a major role in the creation of interactive microworlds to help students learn science (Linn et al., 2006). The ThinkerTools curriculum, designed to assist students in learning Newtonian physics, is an early example that has undergone continuous development since its introduction in 1984. More recent ecological approaches enable students to explore simulated interactions among organisms by creating or discovering simple rules. The importance of visual thinking in chemistry has motivated the construction of software to model three-dimensional chemical structures, animations of chemical reactions, and dynamic linking of the animations with graphs, symbolic expressions, and video. Recent research on teaching students how to design experiments has shown that virtual objects are as effective as real objects in teaching scientific design.

14 Mathematics Instructional Software

Jim Kaput had a dream. He called it "democratizing access to knowledge." His vision was that *every* student could learn algebra if provided with an appropriate curriculum. Kaput did not mince words when describing his disdain for the traditional methods of teaching algebra:

> School algebra in the U.S. is institutionalized as two or more highly redundant courses, isolated from other subject matter, introduced abruptly to post-pubescent students, and often repeated at great cost as remedial mathematics at the post secondary level. Their content has evolved historically into the manipulation of strings of alphanumeric characters guided by various syntactical principles and conventions, occasionally interrupted by "applications" in the form of short problems presented in brief chunks of highly stylized text. All these are carefully organized into small categories of very similar activities that are rehearsed by category before introduction of the next category, when the process is repeated. The net effect is a tragic alienation from mathematics for those who survive this filter and an even more tragic loss of life-opportunity for those who don't. (Kaput, 1995b, p. 71)

Kaput had many ideas for reforming the mathematics curriculum. The most fundamental was to reduce the layer-cake approach that taught mathematics as a sequence of layers—arithmetic, algebra, geometry, precalculus, and calculus. The alternative is a strands organization in which major ideas weave throughout the curriculum, gradually drawing on ever more diverse experiences. These major ideas include quantity, shape, uncertainty, dimensions, and change (Kaput, 1995a).

The mathematics of change particularly interested Kaput because he believed that understanding change is becoming increasingly important for everyone. However, this topic is typically taught only to the relatively few students who take calculus. A democratization of knowledge would need to make this topic available to all students and Kaput argued that technology could make this happen through simulations. It was now possible to break loose from the constraints of the old static media to develop algebra as a tool for representing rates and accumulations of varying quantities.

SIMCALC

Collaborating with Jeremy Roschelle at SRI International, Kaput developed a simulation environment called SimCalc to help middle-school students understand change (Roschelle, Kaput, & Stroup, 2000). The Clown Problem is a typical example.

The Clown Problem

A clown walks at a rate of 3 meters per second for 2 seconds. He then uniformly decreases his velocity over the next 2 seconds from 3 meters to 1 meter per second. He then walks at 1 meter per second for 2 more seconds. How far did he walk?

SimCalc simultaneously produces a simulation and a graph of the clown's progress (Figure 14.1) so students can see the relation between the two representations. The graph makes the problem accessible to middle school students by taking advantage of their ability to count grid squares. Because distance is the product of velocity and time, it can be measured as the area beneath a graph in which velocity is plotted as a function of time. By walking at 3 meters per second over the first 2 seconds, the clown travels 6 meters, which

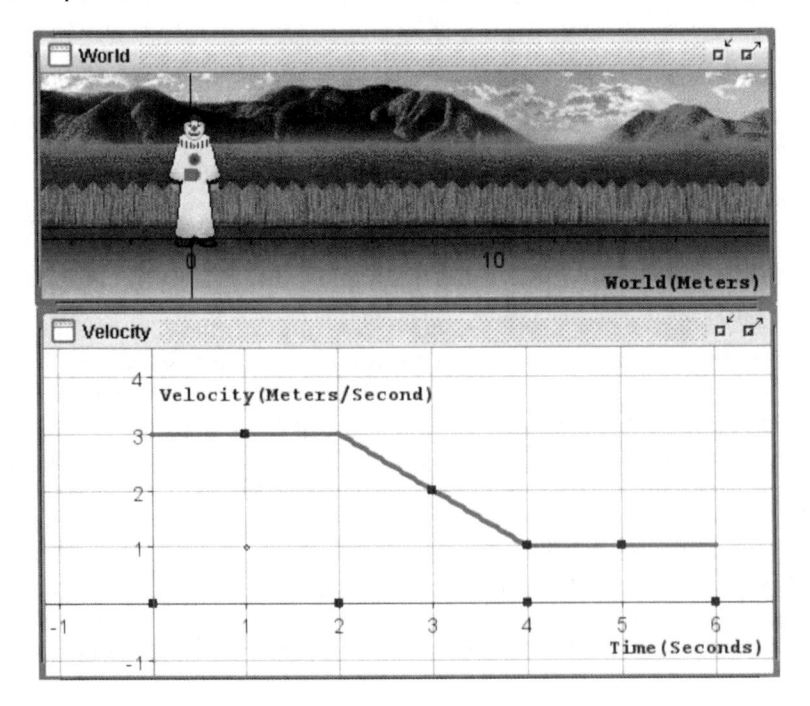

FIGURE 14.1 The clown problem enables students to use area under a graph to find the total distance traveled by the clown. (From "SimCalc: Accelerating Students' Engagement with the Mathematics of Change," by J. Roschelle, J. J. Kaput, and W. Stroup. In *Innovations in Science and Mathematics Education: Advanced Designs for Technologies of Learning* (pp. 48–75), by M. J. Jacobson and R. B. Kozma (Eds.), 2000, Mahwah, NJ: Lawrence Erlbaum Associates. Copyright 2000 by Lawrence Erlbaum Associates. Reprinted with permission.)

is shown spatially by the six squares below the graph. The graph cuts through two squares over the next 2 seconds as the clown slows from 3 meters to 1 meter per second. Adding these two half squares to the other three squares under this section of the graph tells us that the clown traveled an additional 4 meters. The final 2 seconds cover two more meters. The total distance of 6 + 4 + 2 meters can be determined from the total area under the graph. Those of you who have studied calculus will recognize that finding the area under a graph is a fundamental idea in calculus.

Figure 14.2 illustrates another challenge that middle school children can solve by interacting with SimCalc. The goal is to find the constant velocity that the momma fish must swim in order to travel the same distance as the baby fish at the end of 12 seconds. The graph below the simulation shows the erratic behavior of the baby fish, which prefers to swim at different velocities with frequent rest intervals.

SimCalc initially introduces the concept that average velocity can be found by dividing the total distance traveled by total time. The average speed of the clown, for example, would be the 12 meters he traveled divided by 6 seconds. Students

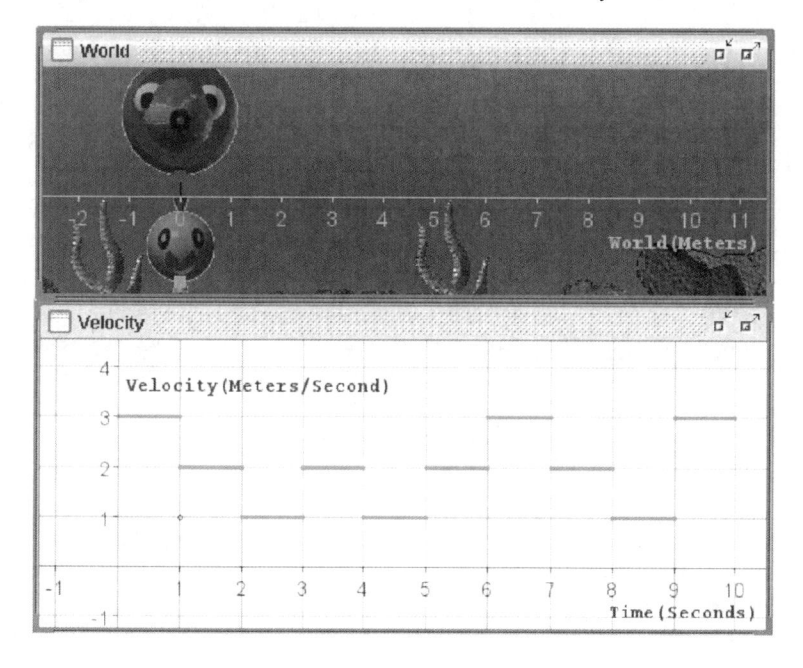

FIGURE 14.2 The fish problem requires students to find a velocity for the momma fish so she will travel the same distance as the baby fish. (From "SimCalc: Accelerating Students' Engagement with the Mathematics of Change," by J. Roschelle, J. J. Kaput, and W. Stroup. In *Innovations in Science and Mathematics Education: Advanced Designs for Technologies of Learning* (pp. 48–75), by M. J. Jacobson and R. B. Kozma (Eds.), 2000, Mahwah, NJ: Lawrence Erlbaum Associates. Copyright 2000 by Lawrence Erlbaum Associates. Reprinted with permission.)

can find the total distance traveled by the baby fish by again counting the number of squares beneath the linear segments of the graph. The baby fish travels 20 meters during the 10 seconds. If the students did not calculate that the momma fish traveled 2 meters per second, the momma fish would be either behind or ahead of the baby fish at the end of the animation. Students would need to go back to the "drawing board" to figure out what they did wrong.

A powerful feature of SimCalc is that it allows students to manipulate one representation to see how it influences another representation. Bowers and Nickerson (2000) took advantage of this feature when they designed a 3-week curriculum on graphs for students in a seventh-grade mathematics class. Bowers and Nickerson began by showing graphs that students had to recreate by acting out the motion. A motion detector recorded and plotted their movement so students could see whether they had correctly reproduced the graph. Other activities required making, evaluating, and defending predictions about mathematics problems by using the simulation capabilities of SimCalc. In a concluding activity, the students created their own simulations such as having two characters meet to exchange secrets, an alien snatching a baby fish, or sports cars racing each other. Performances on a written pre- and posttest indicated that students had improved on many activities that involved interpreting graphs and understanding the relationships among rate, time, and accumulation.

THE ANIMATION TUTOR™

My interest in computer simulations began in the mid-1980s when my research indicated that animated feedback improved students' ability to estimate answers to word problems. My eventual goal was to build on these findings by developing instructional software. This goal receded into the background for a decade before I began to notice the promising simulation-based systems that science educators were creating. However, my most important inspiration was Jim Kaput's vision as implemented in SimCalc.

In 1999, I received a curriculum development grant from the National Science Foundation that allowed me to begin the Animation Tutor™ project. The Animation Tutor™ differs from SimCalc because it is directed toward secondary students and those college students who never mastered algebra during their journey to college. It currently consists of the eight modules shown in the Appendix. Each requires approximately 1 hour to complete.

This book contains a DVD of these modules and I encourage you to explore them. The modules are implementations of the many ideas discussed in this book. The next two sections provide glimpses of the Animation Tutor™ instruction. The first discusses object manipulation and the second discusses simulation. The section on modeling population growth covered in Chapter 9 illustrates parameter variation in another of the modules. All these applications—object manipulation, simulation, and parameter variation—demonstrate how technology can take us well beyond the limitations of static media.

OBJECT MANIPULATION

The Dimensional Thinking module differs from the other modules because it is probably more useful for middle school students than for high school students. My colleague Brian Greer was the principle designer of the module. Greer came to the project with an impressive resume that included an undergraduate degree in mathematics from Cambridge University, a PhD in developmental psychology from Queens University in Belfast, prior experience as the crossword puzzle editor of the *London Times*, and many contributions to the field of mathematics education. However, it was his prior research on people's tendency to overuse proportional reasoning that most interested me.

The Dimensional Thinking module begins with a sign in the window of a pizza parlor that shows the prices of pizzas with different diameters. It then raises the following question.

The Pizza Problem

A 12-inch pizza sells for $6.99 and a 20-inch pizza sells for $12.99. Which is the better value?

Easy to solve! Dividing $6.99 by 12 inches reveals that the first pizza sells for $0.58 per inch and dividing $12.99 by 20 inches reveals that the second pizza sells for $0.65 per inch. The smaller pizza is clearly the better value. We might pause to reflect at this point that the answer is counterintuitive. Doesn't real-world experience teach us that buying more, rather than less, results in better values? Aren't economy sizes usually the larger amounts?

Research conducted by Greer (1993) with 13- and 14-year-old students in Northern Ireland revealed that their answers to math problems usually ignored such real-world knowledge. They applied proportional reasoning to most problems including the following:

1. A man wants to have a rope long enough to stretch between two poles 12 meters apart, but he only has pieces of rope 1.5 meters long. How many of these would he need to tie together?
2. An athlete's best time to run a mile is 4 minutes and 7 seconds. About how long would it take him to run 3 miles?

Typical answers were that the man would need eight pieces of rope to stretch between the two poles and an athlete could run 3 miles in 12 minutes and 21 seconds.

Greer later collaborated with Lieven Verschaffel and Erik de Corte on the book *Making Sense of Word Problems* (Verschaffel, Greer, & de Corte, 2000). The book reviewed research from all over the world that found similar results. People indiscriminately applied proportional reasoning to problems without reflecting on whether it made sense.

One consequence is that many people believe that doubling the diameter of a circle or each side of a square will double its area and that doubling the diameter of a sphere or each side of a cube will double its volume. Even graphic designers make mistakes. You likely will not have to look very hard in magazines or newspapers to find dimensional analysis errors in diagrams. Tufte (2001) provides numerous examples in his chapter on graphical integrity from his book, *The Visual Display of Quantitative Information*. My favorite example on page 70 comes from a 1978 article in the *Washington Post* that displays the diminishing purchasing power of the U.S. dollar. The graphic shows dollars growing smaller and smaller from the 1958 Eisenhower administration through the 1978 Carter administration. The purchasing power of the 1978 dollar had shrunk to 44 cents but its graphic portrayal appeared tiny compared to the 1958 dollar. The reason was that both its length and width were reduced to 0.44 the size of its 1958 counterpart so its area was reduced to $(0.44)^2$ the area of the standard. The designers should have reduced the length and width to the square root of 0.44 of the standard to create an area that was 0.44 the area of the 1958 dollar.

The Dimensional Thinking module allows users to manipulate shapes to see the connections between length, area, and volume. Let us return to the pizza problem. As shown in Figure 14.3, students can drag and superimpose the 12-inch pizza over the 20-inch pizza to discover that the 20-inch pizza is over three times larger than the 12-inch pizza. It is clearly the better buy. Other exercises in the Dimensional Thinking module include determining how many small bowls it would take to fill

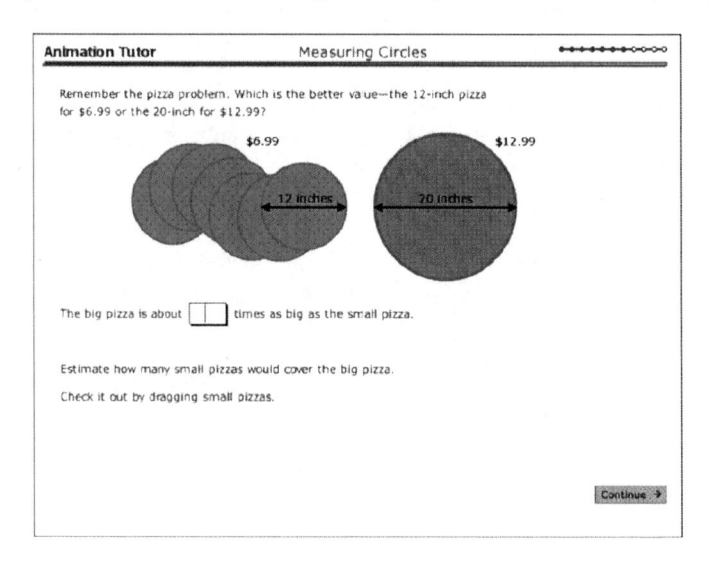

FIGURE 14.3 The Animation Tutor: Dimensional Thinking module allows students to compare the relative areas of two pizzas that differ in diameter. (From Animation Tutor: Dimensional Thinking [Computer software], by B. Greer, B. Hoffman, and S. K. Reed, 2009, San Diego, CA: San Diego State University. Copyright 2009 by San Diego State University. Reprinted with permission.)

a larger bowl that has twice the diameter and how dimensional analysis applies to squares, cubes, and irregular figures. Users learn that doubling each side of a square or the diameter of a circle makes its area four times as large and doubling each side of a cube or the diameter of a sphere makes its volume eight times as large.

SIMULATION

Object manipulation is central to the Dimensional Thinking (and Personal Finance) module but simulation is central to those modules that model dynamic events. A typical example is the initial problem in the Average Speed module.

The Average Speed Problem

Driving east between two cities, a car's speed is 60 mph. On the return trip, it drives 30 mph. What is its average speed?

The simulation of this problem is similar to the fish problem in SimCalc. Users estimate the average speed and then receive animated feedback by watching two cars. One car simulates the two speeds in the problem and the other car simulates the estimate. The two cars (like the two fish) finish the trip at the same time if the estimate is correct. Entering the typical answer—45 mph—produces a simulation in which the car traveling at 45 mph returns before the other car so users learn that the arithmetic average of the two speeds is incorrect. The correct answer is 40 mph because the driver spends twice as much time traveling at 30 mph than traveling at 60 mph:

$$\text{Average speed} = 2/3 \times 30 \text{ mph} + 1/3 \times 60 \text{ mph} = 40 \text{ mph}$$

The Catch Up module also relies extensively on animated instruction. The typical problem describes one vehicle catching another by traveling at a faster speed. The module begins with these types of problems but progresses to more challenging variations such as the Bridge Problem.

The Bridge Problem

A man is standing on a bridge, 300 feet from the left side and 500 feet from the right side. A train is approaching the left side. If the man runs at a speed of 10 mph toward the train, he will reach the left end of the bridge just as the train does. If he runs at a speed of 10 mph away from the train, he will reach the right end of the bridge just as the train overtakes him. What is the speed of the train?

The Bridge Problem can be solved algebraically by constructing two equations to solve for the two unknowns: the location and speed of the train. However, proportional reasoning following a mental simulation that reveals the location of both the man and the train also can solve it. You may want to attempt this approach before reading further.

The simulation approach is described with the aid of an animation that ends in the screen shown in Figure 14.4. It is now possible to use proportional reasoning to find the answer.

Animations become increasingly valuable as problems become more complex, as illustrated by the Braking Problem. The purpose of determining a safe driving distance in this problem is to *avoid* catching up.

The Braking Problem

Imagine that you are driving behind a car at a speed of 24 m/sec. Assume that both you and the driver ahead of you are going the same speed and will decelerate at a rate of 6 m/sec^2 after hitting the brakes. You want to know what the minimal distance is that you can follow the car to avoid hitting it. Assume your reaction time to hit the brakes is .7 sec.

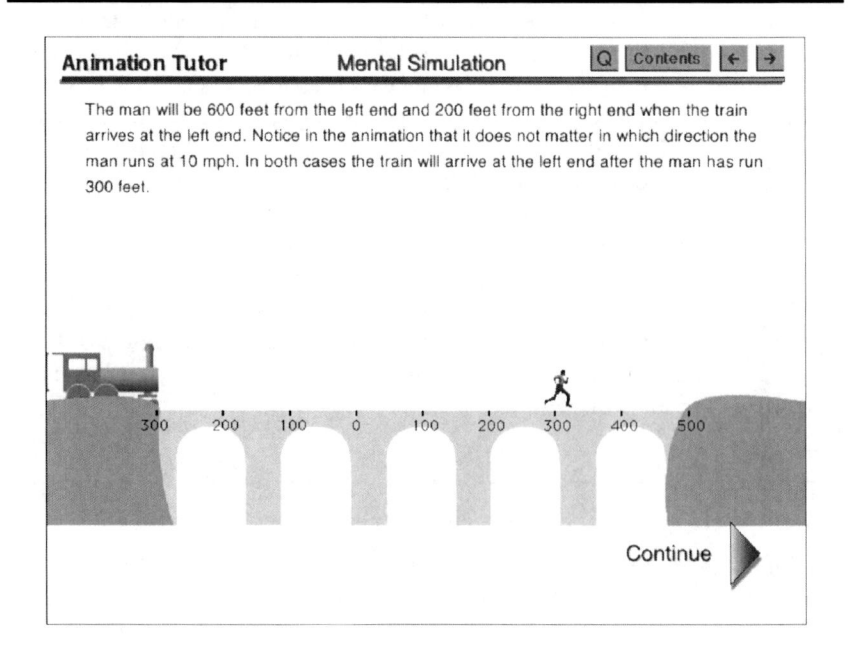

FIGURE 14.4 The Animation Tutor: Catch Up module encourages students to use mental simulation to find the location of both the man and the train in the bridge problem. (From. Animation Tutor: Catch Up [Computer software], by S. K. Reed, B. Hoffman, and S. Phares, 2009. San Diego, CA: San Diego State University.)

The instruction initially asked students to select, among four alternatives, the equation or equations they would need to solve the problem. The four alternatives are

1. Distance = Rate × Time
2. Distance = 1/2 Acceleration × Time^2
3. Both equations 1 and 2
4. Neither equation 1 nor 2

We expected that students in an Intermediate Algebra class would incorrectly include the second equation in their selection because it contains the unknown variable (distance) and two variables (acceleration, time) that have numerical values in the problem. Their selections confirmed our expectations. Equation 2, however, is unnecessary because both cars decelerate at the same rate and, therefore, travel the same distance after braking.

Equation 2 is necessary if the two cars decelerate at different rates, so the instruction next told students to imagine that they were now driving a heavier car that decelerates at a rate of 4 m/sec^2 rather than 6 m/sec^2. The instruction assessed their conceptual understanding by asking them to estimate a safe driving distance for this problem after being reminded of the answer to the first problem. Ten of the 20 college students who answered this question estimated a safe driving distance that was *less* than the safe driving distance for the lighter car! The only explanation that we can suggest for why these students decreased the safe driving distance is that they may have believed the slower rate of deceleration had the same consequences as driving at a slower speed.

We expected that telling them to imagine that they were driving a heavier car would be a clue to increase the distance, but young drivers may lack experience in driving vehicles that differ in weight. Chapter 3 provided many examples of how estimation skill depends on experience. Playing board games helped Head Start children estimate the location of numbers on a number line; using familiar objects to represent standard lengths helped third-grade students estimate longer lengths; and explaining the range, quantity, and linearity principles within a temperature context helped college students estimate the concentration of mixtures.

The advantage of modeling real-world problems in the Animation Tutor™ is that, in addition to learning computational techniques, students receive feedback on their estimates and in the process gain an understanding that may not be provided by experience. Rapid increases in populations and credit card debt because of exponential growth and the need to increase safe driving distance with more weight or slippery pavement are important, practical issues. Creating more modules with these kinds of practical problems is on my wish list if funding enables us to continue the Animation Tutor™ project.

ANIMATE

One of the pieces of Kaput's vision was that technology could free mathematical symbols from their static shackles. Symbols should be easier to comprehend if they could be used to control phenomena. The ANIMATE learning environment designed by Nathan, Kintsch, and Young (1992) at the University of Colorado provides this opportunity.

The purpose of ANIMATE is to provide visual feedback to algebra students by simulating their constructed equations for word problems. Figure 14.5 shows the theoretical assumptions and their implementation for a problem in which a helicopter and train meet by traveling toward each other. The theory assumes that students read the problem to construct a model (mental simulation) of the situation described by the text. Students also construct a problem model that specifies algebraic and mathematical relations in an equation. ANIMATE uses the constructed equation to produce a physical simulation of the problem. The student then compares this physical simulation with her mental simulation to determine if the equation is correct. If the two simulations do not match, the student attempts to repair the equation so the two simulations do match. Nathan et al.'s (1992) research revealed that students who used ANIMATE improved significantly more on a posttest than did students who only constructed mathematical equations without the help of animated feedback.

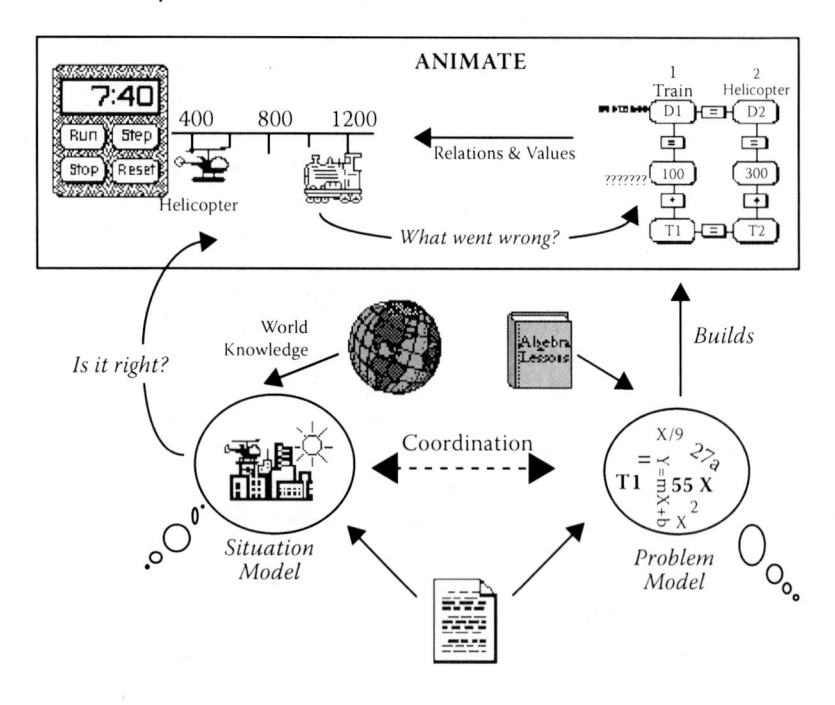

FIGURE 14.5 Theoretical assumptions and their implementation in ANIMATE. (From "A Theory of Algebra-Word-Problem Comprehension and its Implications for the Design of Learning Environments," by M. J. Nathan, W. Kintsch, and E. Young, 1992, *Cognition and Instruction, 9*, 329–389. Copyright 1992 by Lawrence Erlbaum Associates. Reprinted with permission.)

One component of the theory that will probably not survive problems that are more complex is the assumption that students can produce a mental simulation of the problem. We have seen numerous examples in previous chapters in which people understand text by producing mental simulations, so this is a reasonable assumption for simpler problems. However, the bridge and braking problems are challenging because of the difficulty of producing mental simulations that reveal useful spatial relationships. Once these spatial relations are identified, the mathematical procedure for solving each problem is relatively simple. Most of the problems in the Animation Tutor™ modules begin as estimation problems so students receive animated feedback *before* they are asked to construct equations. We need more research to identify which problems require instructional animations to help students correctly construct situation models. SimCalc, the Animation Tutor™, and ANIMATE can all provide such simulations when they are needed.

Roschelle et al. (2000) conclude their article on SimCalc by arguing that technological innovation provides an opportunity to dramatically restructure school curricula to support visualization, simulation, and modeling that are becoming increasingly important in a technological world. Preliminary results from scaling-up SimCalc for a 3-week replacement unit on understanding proportionality through functioning demonstrates that this promise can be fulfilled. A completed pilot experiment involving 21 seventh-grade math teachers from Texas revealed that innovative technologies can have an important impact on student learning (Tater et al., 2008). Jim Kaput would be delighted.

SUMMARY

Kaput advocated the use of technology as the basis for the reform of an algebra curriculum that would begin in middle school by including topics such as rate and accumulation. The goal of his SimCalc project is to make algebra accessible to all students by dynamically linking animations with graphs of functions. The Animation Tutor™ project has similar goals but is designed for high school and college algebra classes. It enables students to view an animation of the problem to test the accuracy of their estimates and calculations. ANIMATE is another animation-learning environment in which students evaluate whether their constructed equations are correct by determining whether the equation produces a correct animation of the problem. Preliminary findings from scaling up SimCalc suggest that innovative technologies can have an important impact on student learning.

15 Conclusions

Let me begin by reviewing the main points of this book before discussing their implications. Over the past decade, there has been increasing interest in embodied cognition in which perception and action play a central role in thinking. Barsalou's (1999) perceptual symbols hypothesis has contributed to this movement by arguing that mental simulations reenact our initial perceptual experiences and therefore thinking is not solely based on abstract, amodal knowledge. Many experimental findings support the view that much of thinking can be viewed as reliving our perceptual and motor experiences, rather than as a form of internal verbalization.

In babies, the visual system lays the foundation for objects and concepts that can later be described in words. It creates discriminable patterns such as faces, categories such as animals and vehicles, and image schemas such as containment and path. The visual system also provides an approximate sense of large quantities and an exact count of small quantities. Spatial skills provide the foundation for estimating answers in a variety of measurement tasks that include building a mental number line in which numbers are equally spaced, using familiar objects to estimate length, and interpolating between numbers in proportional reasoning tasks. Mathematics instruction should include exercises for developing these spatial skills.

Spatial metaphors and the creation of visual images also support thinking. Many concepts are based on metaphors and some of these, such as the association of consciousness, health, control, goodness, and virtue with *up*, are spatial. Even abstract domains such as mathematics rely on spatial metaphors as simple as paths and containers. Although metaphors typically work unconsciously, visual images are conscious creations that help us learn and reason. The manipulation of images is useful in a variety of problem-solving tasks that involve mental rotation, causal reasoning, and the creation of products from components. Fortunately, our visual imagery is usually good enough for improving memory and thinking without being so good that it creates a breakdown in reality monitoring.

Understanding visual displays such as pictures, diagrams, and graphs is crucial in many professions. There is much evidence that static pictures aid learning when people attend to the relevant parts. Animation can direct attention, but its effectiveness varies greatly across tasks. The challenge for researchers is to identify when instructional animations are required because mental simulations are inadequate. Diagrams are abstract visual representations that preserve important spatial relations. The skillful use of diagrams such as matrices, networks, hierarchies, and Venn diagrams requires knowledge of how they differ in representing these relations. Comprehending graphs requires skills in reading data, finding relationships in data, and drawing conclusions from data. Understanding rates of change is particularly challenging. This is unfortunate because comprehending the consequences of exponential growth, including population growth, is limited by underestimation of its explosiveness.

Much of learning depends on our ability to integrate words and pictures. A variety of research paradigms have provided evidence that visual simulations support comprehension of spatial relations in text including the generation of predictive and casual inferences. Vision also combines with action to improve both memory and reasoning. For example, pantomiming verbal phrases increases recall relative to saying the phrases aloud. As emphasized by Piaget (1977), however, actions must be accompanied by reflective thinking to be effective. Both vision and action are emphasized in virtual environments that create physical simulations of real-world activities. Theories of these activities, such as Neisser's (1967) perceptual cycle, describe a more cyclical interplay between perception and action than is emphasized in models that are more traditional. One example is Klein's (1993) recognition-primed model in which situation awareness is the key component of making good decisions in risky environments. Training and experience increase situation awareness, as illustrated by the training of novice drivers and the evaluation of electronic warfare technicians.

MULTIMEDIA LEARNING

The interplay between vision and action that occurs in virtual reality training can also occur in multimedia classroom instruction. Animation, for example, has played a major role in the creation of interactive microworlds to help students learn science. Examples include the ThinkerTools curriculum that assists students in learning Newtonian physics, ecological software that enables students to explore simulated interactions among organisms, chemistry programs that model three-dimensional chemical structures and chemical reactions, and experimental design instruction that substitutes virtual objects for real objects. Examples of animation-based mathematics programs include SimCalc to make algebra accessible to middle school students by dynamically linking animations with graphs of functions, the Animation Tutor™ to provide animated feedback on estimated and calculated answers, and ANIMATE to provide feedback on whether constructed equations produce a correct animation of the problem.

All of these virtual reality and multimedia projects require extensive evaluation to determine their effectiveness (Baker et al., 2008). This need is becoming critical as computers become more widely used in delivering instruction. In July 2006, the University of California announced that it was opening a virtual public high school for San Diego County students. This is not an unusual occurrence. Virtual schools are opening in many cities across the country. Clark (2005) informs us that, although it is still distant to traditional classroom instruction, computer-based instruction is steadily increasing. By the year 2000, almost 90% of universities with more than 10,000 students offered some form of distance learning. Computers delivered 16% of instruction in American business and industry in 2003, up from 12% the previous year. Industry's total annual investment in training at that time was between $50 billion and $60 billion. Electronic

Conclusions 175

learning provided cost savings by decreasing travel costs and reducing instructional time.

However, why is this important? Isn't our current educational system doing an adequate job of educating students? Numerous reports agree that the answer is clearly no.

THE NEED

In 1983, the U.S. Department of Education National Commission on Excellence in Education (1983) published the report *A Nation at Risk* that expressed alarm at the rising mediocrity in education. This alarm was echoed in another report issued during the same year by a commission of the National Science Board (1983). Both reports set the goal that American precollege students would become the best in the world in mathematics and science. What has happened since the publication of those reports? According to the National Science Board's more recent assessment:

> In the intervening years, we have failed to raise the achievement of U.S. students commensurate with the goal articulated by that Commission—that U.S. precollege achievement should be "best in the world by 1995"—and many other countries have surpassed us. Not only are they not first, but by the time they reach their senior year, even the most advanced U.S. students perform at or near the bottom on international assessments. (2006, p. 1)

This statement, from *America's Pressing Challenge—Building a Stronger Foundation,* again sounds an alarm that rings even louder today than it did in 1983. Science and engineering occupations have grown at more than four times the annual growth rate of all occupations since 1980. Changes in the workforce require that new workers need increasingly more sophisticated skills in science, mathematics, and technology. The report warns that the United States must better prepare its students if it is to maintain its economic leadership and compete in the new global economy.

The National Science Board makes a number of specific recommendations, one of which is to take advantage of the opportunities offered by new communication technologies. The challenge for K–12 education is to shift professional development of teachers away from just learning to use the computer toward more effective use in supporting instructional goals. The report recommended simulations, specialized laboratories, Web-based research, data collection, analysis projects based outside the school, and communication with experts and peers.

This recommendation is consistent with one made several years earlier in the U.S. Department of Commerce's *Visions 2020: Transforming Education and Training through Advanced Technologies.* According to its follow-up report *Visions 2020.2*:

Advances in both cognitive science and information technology have the potential to transform education and training in ways previously unimaginable. Advance technologies under development by U.S. businesses, universities, and government could create rich and compelling learning opportunities that meet all learners' needs, and provide education and training when and where they are needed, while boosting the productivity of learning and lowering its costs. These technologies could play a major role in meeting education and training challenges in the years ahead, and help make the U.S. workforce more competitive globally. (U.S. Department of Commerce, 2005, p. 4)

In an interesting departure from most reports, *Visions 2020.2* also describes students' views. During October and November 2004, NetDay asked K–12 students across the country how they could benefit from new technologies. More than 160,000 replied to online questionnaires. The Commerce Department reviewed 8,000 of these responses for common themes and formulated a composite of the most popular suggestions:

Every student would use a small, handheld wireless computer that is voice activated. The computer would offer high-speed access to a kid-friendly Internet, populated with websites that are safe, designed specifically for use by students, with no pop-up ads. Using this device, students would complete most of their in-school work and homework, as well as take online classes both at school and at home. Students would use the small computer to play mathematics learning games and read interactive e-textbooks. In completing their schoolwork, students would work closely and routinely with an intelligent digital tutor, and tap a knowledge utility to obtain factual answers to questions they pose. In their history studies, students could participate in 3-D virtual reality-based historic reenactments. (U.S. Department of Commerce, 2005, p. 6)

These reports identify a tremendous need to improve instruction. They also indicate that advanced technologies provide a potential solution. Is this likely to occur in the near future?

THE FUTURE?

Each year a group of experts in the use of technology on college campuses describes six areas of emerging technology that will likely have a significant impact on higher education within the next 5 years. For the first time in the 2006 edition of *The Horizon Report* (Johnson & Smith, 2006) this group identified critical challenges for implementing these technologies:

1. The development and evaluation of instructional software has typically been carried out by university faculty. Yet the promotion and tenure of these faculty have traditionally been based on print-based publications rather than on the construction of software.

Conclusions

2. Although most of today's students are surrounded by technology, they have not used this technology to develop critical thinking skills.
3. The management of intellectual property rights regarding digital media has not been sufficiently addressed.
4. The demonstration of success of new technologies does not quickly scale up for widespread usage.
5. New technologies require support and the willing participation of people who are asked to change the way they work.

These challenges, combined with constraints on funding the construction and evaluation of instructional software, raise some difficult but important questions. Will the United States have a different system of education in the year 2020? Will this system use technology to support thinking by providing interactive learning environments in which perception and action provide a foundation for formulating and testing ideas?

Let me try to answer these questions through a story that links the components of embodied cognition (seeing, acting, and thinking) with our conception of the future. Our body provides a reference for our conception of time. The future is in front of us and the past is behind—except for the Aymara Indians in the mountains of northern Chile.

Rafael Nuñez, whose work with George Lakoff was discussed in Chapter 4, and Eve Sweetser at the University of California, Berkeley discovered that the words and gestures of the Amarya describe how the past lies in front of them and the future lies behind (Nuñez & Sweetser, 2006). The Amarya refer to the past by sweeping their hands in front of their body and using a word that means "front" or "sight." They appear to metaphorically place the past in front of them because only the past is knowable. The future lies behind them, sight unseen.

Like the Amarya, I can only see the past. My prediction of the future therefore relies on my current observations:

- Behavioral and neurological research has demonstrated that perception and action are not peripheral, but central components of thinking.
- Multimedia instruction can build on these findings by providing interactive learning environments in which perception and action support learning.
- Research-proven instruction is becoming increasingly important as electronic learning plays a more central role in education and training.
- The need for this instruction in the United States is demonstrated by the failure of American students to rank high on international mathematics and science exams.
- The lack of government funding to support research and development of instructional software in mathematics and science education will assure that multimedia instruction will have little impact on education.

Will we have an educational system in the year 2020 that is supportive of visual thinking? It may happen in other countries but I doubt it will happen in the United States. I cannot find a commitment for making it happen. I hope I am wrong.

Appendix: The Animation Tutor™ DVD

Stephen K. Reed and Bob Hoffman

The Animation Tutor™ DVD is a curriculum development project that was created through a grant from the National Science Foundation. It contains eight modules that encourage and support visual thinking through manipulation and animation to help people reason about mathematics. The mathematical content in these modules is typically taught in high school but, because students often struggle with this material, the modules can later be used in college or at home as a review of content that many students never learned.

Each module provides instruction on a particular situation, such as population growth, rather than on a particular mathematical procedure such as exponential functions. One advantage is that users can more easily compare and contrast different mathematical content such as linear and exponential models of population growth. A second advantage is that focusing on situations emphasizes the application of mathematics. A third advantage is that the modules can also be used in other courses that include the topics. A course on biology could use the Population Growth module, a course on finance could use the Personal Finance module, a course on chemistry could use the Chemical Kinetics module, and a course on physics could use the Catch Up module on acceleration.

Many of the problems initially require estimates to improve estimation skills and provide motivation for learning analytic techniques for calculating the answers. Animation of students' answers (such as the time to fill a tank or complete a round trip) provides visual feedback on the accuracy of their estimates and calculations. Some problems also include graphs so students can attempt to infer what kind of functional relations link the variables.

The following provides an overview of the eight modules. More information about their use is contained on the DVD.

DIMENSIONAL THINKING
Brian Greer, Bob Hoffman, and Stephen Reed

The Dimensional Thinking module begins with the question of whether a 12-inch pizza for $6.99 or a 20-inch pizza for $12.99 is the better buy. Dividing price by diameter results in the incorrect answer that the smaller pizza is the better buy. The purpose of this and other problems is to correct the overuse of proportional reasoning to situations in which it does not apply. The module demonstrates that doubling the side of a square or the diameter of the circle doubles its perimeter but does not double its area. Students manipulate geometric objects to learn that doubling a side of a square or the diameter of a circle increases area by a factor of four. Doubling the side of a cube or the diameter of a sphere increases volume by a factor of eight. Applications to standard and irregular forms show both correct (perimeter) and incorrect (area, volume) uses of proportional reasoning.

CHEMICAL KINETICS

Kathy Tyner, Stephen Reed, and Susan Phares

The Chemical Kinetics module uses the context of chemical reactions to discuss two concepts that are important in calculus—area under a curve and the tangent to a curve. Students estimate the proportion of a curve that exceeds a critical value of kinetic energy to estimate the proportion of atoms that will form molecules. They make these predictions for kinetic energy distributions at low and high temperatures before viewing simulations. The concepts of average and instantaneous rates of change are illustrated for a curve showing the half-life of aspirin. The instruction shows that it is possible to find an instantaneous slope (represented by a tangent) that has the same value as the average slope (represented by a secant) for different intervals of the curve.

PERSONAL FINANCE

Bob Hoffman and Stephen Reed

The Personal Finance module compares investing money at simple and compound rates of interest to illustrate the difference between linear and exponential growth. Students calculate how their money grows over a 5-year period for simple and compound rates of interest. They also use algebra to determine how much money they would earn by partitioning their savings between a certificate of deposit and a savings account. The mathematics of investing are also applied to borrowing money, such as calculating monthly payments on a $200,000 mortgage at 6.5% interest over 15 years and the same loan at 7% interest over 30 years. Graphs of the remaining principle for loans and the amount earned for investments depict how the amount of money changes over time.

POPULATION GROWTH

Stephen Reed, Bob Hoffman, and Diane Short

The Population Growth module compares linear and exponential models of population growth by computing a (least squares) goodness-of-fit index between actual and predicted data. Students adjust the y-intercept and slope to try to find the best-fitting linear model of population growth in the United States during the 19th century. They then compare this model to an exponential model of the same data. Students next make predictions about the accuracy of extrapolating the best-fitting linear and exponential models to predict population increases in the 20th century. The distinction between linear and exponential growth is illustrated by asking when a tank of bacteria will be one-quarter filled if the bacteria double every minute beginning at 11:00 am and fill the tank at noon. An animation and simultaneous graphing of the increase in population provides feedback about the explosiveness of exponential growth. The instruction concludes by applying polynomial functions to world population growth.

AVERAGE SPEED

Stephen Reed, Jeff Sale, and Susan Phares

The Average Speed module provides instruction about weighted averages by showing that average speed is a weighted average of the amount of time spent traveling at different speeds. Students begin by estimating the average speed of a round trip for a speed of 60 mph on the initial trip and 30 mph on the return trip. Animated feedback reveals that 45 mph is incorrect because a car traveling at this speed in both directions returns sooner than the other car. Students continue to receive animated feedback as they practice estimating average speeds for different initial and return speeds. The counterintuitive idea that the average speed can never exceed twice the slower speed is explored *graphically* as the asymptote of a function, *conceptually* as total distance divided by total time, and *algebraically* as a derivation based on the weighted average formula.

CATCH UP

Stephen Reed and Bob Hoffman

The Catch Up module contains problems in which one person catches another person by traveling at either a constant speed or a constant rate of acceleration. The problems require a variety of mathematical skills including estimating answers, selecting an appropriate equation or mathematical relation, solving single and simultaneous equations, matching graphs to situations, providing conceptual explanations, and planning solutions. The bridge problem shows how a challenging mathematical problem can be easily solved by combining a spatial simulation with proportional reasoning. Determining safe driving distances provides an important context for spatial reasoning about deceleration.

TASK COMPLETION

Stephen Reed, Susan Phares, and Jeff Sale

The Task Completion module provides instruction on solving algebra word problems by showing how the same equation applies to different situations. The situations require finding the time it takes for two pipes to fill a tank, two workers to complete a task, or two people to meet by traveling toward each other. The key to solving these problems is to convert completion times into rates that specify how much of the task is completed during each unit of time. Students also learn how to solve a more complex version of these problems in which two pipes begin filling the tank at different times. In addition to showing solutions based on a single algebraic equation, the instruction shows solutions that decompose a problem into its parts.

LEAKY TANKS
Stephen Reed, Susan Phares, and Jeff Sale

The Leaky Tanks module continues with instruction on more complex variations of the tank-filling problems by introducing problems in which there is a loss of liquid from the tank as it fills. Students are asked and receive feedback on how they would modify the simpler equation to solve these more complex problems. The solution to the bottom-leak problem requires reducing the rate of fill by subtracting the rate of loss. Calculating how long it takes to fill a tank with a side leak can be accomplished by breaking the problem into two parts. The fill time below the leak is found by using the equation for a no-leak problem and the fill time above the leak is found by using the equation for a bottom-leak problem.

Encapsulated Summaries

CHAPTER 1: IMAGES VERSUS WORDS

Both words and images are symbols. According to DeLoache (2004), a symbol is something that someone intends to represent something other than itself. Symbols play an important role in many theories of knowledge representation that have been developed by theorists working in psychology (Markman, 1999; Palmer, 1978) and in artificial intelligence (Davis, 1998; Davis, Shrobe, & Szolovits, 1993). Symbols can also form the basis for intriguing stories, as illustrated by Dan Brown's (2003) novel *The Da Vinci Code*.

Words play an important role in cultural transmission (Gleitman & Papafragou, 2005; Vygotsky, 1962) and in some forms of complex thought (Sternberg & Ben-Zeev, 2001). However, the importance of language in communication creates an overemphasis of its role in thinking (Pinker, 1994). Visual thinking plays a major role in reasoning (Schwartz & Heiser, 2006; Tversky, 2005), including scientific discoveries (Miller, 1984; Shepard, 1988). Visualization is also encouraged in the brilliantly descriptive and metaphoric writing of authors such as Pat Conroy (1986), Truman Capote (1965), and Kaye Gibbons (2006).

The important role of perceptual processing in supporting many cognitive activities can be seen in Barsalou's (Barsalou, 1999; Barsalou, Simmons, Barbey, & Wilson, 2003) articles on perceptual symbols systems and in research on embodied cognition (Gibbs, 2006; Rubin, 2007; Sumin & Smith, 2008; Wilson, 2002). Studies showing that verification times depend on the vertical alignment of words (Zwaan & Yaxley, 2003) and the orientation of objects (Stanfield & Zwaan, 2001) are consistent with Barsalou's perceptual simulation hypothesis. Other support comes from neuroimaging studies (Chen et al., 1998; Hesslow, 2002) that show activation of the visual cortex during thinking.

CHAPTER 2: IMAGES BEFORE WORDS

Fantz's *Scientific American* article on the origins of form perception is a very readable introduction to this topic (Fantz, 1961). Fagan (1973) reports several experiments on infants' delayed recognition memory. There is also research on infants' abilities to generalize over different perspectives of faces (Cornell, 1974; Turati, Bulf, & Simion, 2008).

Jean Mandler's book *The Foundations of Mind: Origins of Conceptual Thought* discusses a theory of image schemas that form the basis of concepts (Mandler, 2004). Much of her theory is based on her own research, such as that reported by Mandler and McDonough (1996). It has also been influenced by the writings of cognitive linguists who emphasize the embodied nature of thought (Johnson, 1987; Lakoff, 1987). Their ideas are presented in Chapter 4. Chatterjee

(2001) expresses a related view—that language and spatial representations converge at an abstract level of concepts and simple spatial schemas.

Stanilas Dehaene's book *The Number Sense: How the Mind Creates Mathematics* (Dehaene, 1997) is a very readable introduction to the ideas presented in the remainder of this chapter. The research on monkeys' and infants' abilities to keep track of small numbers is reported in Hauser (2000) and Feigenson, Carey, and Hauser (2002). There are several very good (but densely written) summaries of the research on nonverbal numerical cognition (Feigenson, Dehaene, & Spelke, 2004; Gallistel & Gelman, 2000; Gelman & Gallistel, 2004).

The finding that different kinds of magnitude judgments depend on the same area of the brain suggests that there may be a common metric for time, space, and quantity (Dehaene, Molko, Cohen, & Wilson, 2004; Walsh, 2003), including evidence that space helps us think about time (Casanto & Broditsky, 2008) and time is linked to magnitude (Kiesel & Vierck, 2009). The initial discovery of the time to decide which of two numbers is larger was made by Moyer and Landauer (1967). Weise (2003) discusses the transition from nonverbal to verbal numerical cognition. The distinction between nonverbal and verbal numerical reasoning is also shown in the research on exact versus approximate addition (Dehaene, Spelke, Pinel, Stanescu, & Tsivkin, 1999).

CHAPTER 3: ESTIMATION

Research by Siegler and Opfer (2003) shows that a logarithmic function proposed by Fechner accurately models how young children initially locate numbers on a number line. The correct linear function is gradually acquired over grades K–4 for the 1–100 and 1–1000 lines. Instruction on moving pieces in board games helps Head Start children improve on the 1–10 number line (Siegler & Ramani, 2006). Children who have learned more accurate number lines also perform better on tasks that require numerosity, measurement, and computational estimation (Booth & Siegler, 2006).

Differences between computational and measurement estimation have been identified by Hogan and Brezinski (2003) using a psychometric approach. They propose that measurement estimation depends on spatial skills that should be encouraged in mathematics courses. Sowder's (1992) chapter on estimation and number sense contains an extensive literature review of earlier work on this topic. Relating standard units of measurement to familiar objects has proven helpful for teaching measurement estimation to children (Joram, Gabriele, Bertheau, Gelman, & Subrahmanyam, 2005; Joram, Subrahmanyam, & Gelman, 1998).

The concept of a mental number line is useful for understanding the range and quantity principles that support the interpolation method for estimating answers to mathematics problems. Estimated temperatures of mixtures can be more accurate than calculated answers (Dixon & Moore, 1996) and help students select better strategies for calculating answers (Ahl, Moore, & Dixon, 1992; Dixon & Moore, 1996). Interpolation is the primary strategy used by college students to estimate

Encapsulated Summaries 185

the temperature of mixtures (Reed, 1999). Instruction on the range, quantity, and linearity principles within the temperature domain is very helpful for improving interpolation strategies within an unfamiliar acid domain (Reed & Evans, 1987).

CHAPTER 4: SPATIAL METAPHORS

The book *Metaphors We Live By* (Lakoff & Johnson, 1980) demonstrated how much conceptual thinking depends on concrete metaphors. For instance, cognitive psychologists have proposed that information flows through a sequence of stages in either a bottom–up or a top–down direction. Neisser's (1967) book *Cognitive Psychology* helped launch the cognitive revolution and influenced my own textbook on cognitive psychology (Reed, 1982, 2006). The Lakoff and Nuñez analysis of metaphors that support mathematical reasoning is described in *Where Mathematics Comes From* (Lakoff & Nuñez, 2000) and in an earlier book chapter (Lakoff & Nuñez, 1997). Some instructional implications of this analysis are discussed by Nuñez, Edwards, and Matos (1999).

The metaphor of traveling along a path applies to inanimate objects through fictive motion (Matlock, 2004) and to arithmetic through stepping out addition and subtraction on a number line (Cemen, 1993). The source–path–goal schema is utilized in cognitive scientists' representation of search spaces for problem solving (Newell & Simon, 1972; Simon & Reed, 1976). Representations of similarity as distances between points in a multidimensional space is an example of a linking metaphor that was created by Shepard's (1957) work on multidimensional scaling. The depiction of similarity relations among birds is a good example of the theoretical application of multidimensional scaling (Rips, Shoben, & Smith, 1973; Smith, Rips, & Shoben, 1974).

CHAPTER 5: PRODUCING IMAGES

Pani (1996) traces the history of the functionalists' interest in imagery, including their influence on contemporary thinking. Watson's (1924) book *Behaviorism* put a damper on the study of imagery for almost 45 years. Heidbreder (1961) describes functionalism, behaviorism, and Gestalt psychology (among others) in her book *Seven Psychologies*.

Paivio (1969) reviews his early research on imagery and learning in a *Psychological Review* article. His later book (Paivio, 1986) on mental representations provides a more extended discussion of dual coding theory. *The Memory Book* (Lorayne & Lucas, 1974) describes techniques for improving memory, many of which utilize visual imagery. Research on an imagery strategy for learning names (Morris, Jones, & Hampson, 1978) is one of many studies on the effectiveness of imagery for learning.

The initial debates on the importance of visual imagery as a theoretical construct can be seen in articles by Pylyshyn (1973) and by Kosslyn and Pomerantz (1977). The debate continues (Kosslyn, Thompson, & Ganis, 2006; Pylyshyn, 2003). The functionalist approach, in contrast, emphasizes the practical applications of imagery such as reducing interference from verbal codes (Brooks, 1968), simultaneously comparing features of patterns (Nielsen & Smith, 1973; Smith & Nielsen, 1970), and creating category prototoypes (Reed, 1972; Reed & Friedman, 1973).

It is difficult, however, to discover new information in images as indicated by poor performances in finding parts (Reed, 1974) and in reinterpreting ambiguous figures (Chambers & Reisberg, 1985, 1992). The overshadowing of visual memories by more impoverished verbal descriptions (Schooler & Engstler-Schooler, 1990) provides a possible explanation of these limitations (Brandimonte & Gerbino, 1993).

Johnson and Raye (1981) propose a theory of reality monitoring based on their previous research (Johnson, Raye, Wang, & Taylor, 1979). Bentall (1990) mentions this research in his review of studies on hallucinations. One study found that schizophrenics report more vivid imagery than do normal people (Sack, van de Ven, Etschenberg, & Linden, 2005). Interest in reality monitoring has led cognitive psychologists to examine clinical procedures used to uncover memories of childhood sexual abuse (Lindsay & Read, 1994; Loftus, 1993; Singer, 1990). The need for care in using imagery to recover memories is demonstrated by experimentally induced false memories (Hyman & Pentland, 1996).

CHAPTER 6: MANIPULATING IMAGES

One of the early studies of performing cognitive operations on visual images investigated whether people scan images in the same way they scan perceptual patterns (Kosslyn, Ball, & Reiser, 1978). Pylyshyn (1981) challenged an imaged-based interpretation of mental scanning by arguing that people use their knowledge of how distance influences time to produce the results. Although people's predictions of their scanning times did produce the expected linear increase in time with distance (Mitchell & Richman, 1980), they were unable to predict how different shapes would influence scanning times (Reed, Hock, & Lockhead, 1983).

Evidence for the manipulation of images has been found for a variety of tasks (Hegarty, 2004) including the mental rotation of complex objects (Shepard & Metzler, 1971) and causal reasoning in pulley problems (Hegarty, 1992), and in gear problems (Schwartz & Black, 1996). Temple Grandin's visual thinking is described in her book *Animals in Translation* (Grandin & Johnson, 2005). Autistics, in general, show much better performance on spatial tasks than on verbal tasks (Dawson, Soulieres, Gernsbacher, & Mottron, 2007).

The Gestalt approach is illustrated by research on reorganization (Kohler, 1925), productive thinking (Wertheimer, 1959), insight (Metcalf, 1986), and self-imposed constraints (Knoblich, Ohlsson, Haider, & Rhenius, 1999). Research on creative cognition is described in the book *Creative Imagery: Discoveries and Inventions in Visualization* (Finke, 1990). Finke, Ward, and Smith (1992) discuss the Geneplore model in their book *Creative Cognition: Theory, Research, and Applications*. Attempts to encourage creativity at Nissan Design America are described in the book *The Creative Priority* (Hirshberg, 1998).

Encapsulated Summaries 187

CHAPTER 7: VIEWING PICTURES

Both pleasant and unpleasant pictures initially attract more attention than unemotional pictures (Calvo & Lang, 2004). The emotional reaction to pictures is illustrated by the responses to caricatures of President Nixon during the Watergate hearings (Wheeler & Reed, 1975). Decorative pictures can be distracting and reduce recall of explanatory text (Harp & Mayer, 1997; Levin, 1989). They may, however, motivate people to read articles by triggering situational interest (Hidi & Renninger, 2006). Research is needed to evaluate the impact of particularly evocative pictures such as the one of the Afghan girl on the June 1985 cover of *National Geographic* (Hajela, 2005).

An extensive review of research indicates that static pictures typically do support learning (Anglin, Vaez, & Cunningham, 2003) although individual differences in students' prior knowledge, visuospatial ability, and strategies influence their effectiveness (Vekiri, 2002). A challenge for instructors is that novices do not always attend to the relevant information in a picture (Lobato, 1996, 2008). Grant and Spivey's (2003) study of students' eye movements during work on Duncker's radiation problem allowed them to increase the number of correct solutions by identifying and highlighting the most relevant information. Viewing pictures is also involved in producing pictures when learners compare their drawings to the correct drawings (Van Meter, 2001; Van Meter & Garner, 2005).

There has been less research on animated pictures and the results are more mixed (Anglin et al., 2003; Rieber, 1990; Tversky, Morrison, & Betrancourt, 2002), but much has been recently learned about building effective animation (Moreno & Mayer, 2007; Plass, Homer, & Haywood, 2009). Animation is particularly effective when it is representational, realistic, and illustrates procedural-motor knowledge (Hoffler & Leutner, 2007). Building animation is not necessary if students can mentally simulate casual events from static pictures (Hegarty, Kriz, & Cate, 2003). However, it is difficult to mentally simulate temporal events at a high level of accuracy so instructional animations have been effective in providing visual feedback (Reed, 2005; Reed & Hoffman, 2004).

CHAPTER 8: PRODUCING DIAGRAMS

A description of Mark Harrower's cartography class appears in the Winter 2005 issue of *On Wisconsin*. The passage from Lewis Carroll is included in Norretranders' (1998) book *The User Illusion*. Both Diezmann and English (2001) and Hegarty and Kozhevnikov (1999) discuss the distinction between schematic and pictorial representations. An excellent overview of representational issues in mathematics and science learning is provided by diSessa (2004).

Carroll, Thomas, and Malhotra (1980) found that students spontaneously used a matrix for designing a floor plan and could effectively use a provided matrix to design a manufacturing process. Research has shown that undergraduates have some general knowledge of selecting diagrams for problems (Novick, Hurley, &

Francis, 1999), although computer science and mathematics education majors have a more detailed understanding of the formal differences that distinguish hierarchies, matrices, and networks (Novick & Hurley, 2001). Research continues on criteria used by college students to select diagrams for solving problems (Hurley & Novick, 2006; Novick, 2006)

The use of semantic networks to model the organization of concepts in long-term memory is illustrated by the Spreading Activation model (Collins & Loftus, 1975). The instructional benefits of constructing semantic networks are discussed by Holley and Dansereau (1984), Gorodetsky and Fisher (1996), and O'Donnell, Dansereau, and Hall (2002). An instructional version of the spreading activation model uses animation to illustrate causal connections among concepts (Schwartz, Blair, Biswas, Leeawong, & Davis, 2008). Embedding pictures within semantic networks (Gorodetsky & Fisher, 1996) offers a promising method for integrating modal and amodal symbols. As effectively argued by Dove (2009), amodal symbols enable us to acquire semantic content that goes beyond perceptual experience.

CHAPTER 9: COMPREHENDING GRAPHS

Tufte's classic book, *The Visual Display of Quantitative Information*, was initially published in 1983. The second edition (Tufte, 2001) appeared 18 years later. Tufte discusses this and his other books in a 2004 interview (Zachry & Thralls, 2004). The data shown in Figure 9.2 are based on research by Doll (1955) that was included in the 1964 Report of the Advisory Committee to the Surgeon General titled *Smoking and Health* (1964). Friel, Curcio, and Bright's (2001) literature review organizes making sense of graphs according to the skills of reading data, finding relationships in the data, and moving beyond the data. Making sense of graphs can be challenging even for experts (Trickett & Trafton, 2007).

Comprehending graphs, as found for producing diagrams, typically depends on attending to spatial relationships rather than pictorial details (Kozhevnikov, Hegarty, & Mayer, 2002). A particularly challenging task requires understanding how to translate a graph of rates of change into a graph of change (Carlson, Jacobs, Coe, Larsen, & Hsu, 2002). The Algebra Sketchpad (Yerushalmy, 1997) enables students to represent sequential changes in quantity by linear and curvilinear icons to help them model rates of change.

Understanding the exponential function is particularly important (Cole, 1998) because of its explosive growth (Philipp, Martin, & Richgels, 1993) and because of the tendency to greatly underestimate this growth (Wagenaar & Sagaria, 1975). The Population Growth module (Reed, Hoffman, & Short, 2009) in the enclosed Animation Tutor™ DVD (Reed & Hoffman, 2009) enables students to create linear and exponential functions to model population growth in the United States between 1800 and 2000.

CHAPTER 10: WORDS AND PICTURES

Baddeley and Hitch (1974) proposed a model of working memory that contained a phonological loop for auditory information and a visuospatial sketchpad for visual information. Baddeley (1992) and Saariluoma (1992) discuss the importance of visual coding in chess. The revised model (Baddeley, 2000, 2001) added a component for integrating both sources of visual and verbal information. Forming spatial mental models from route descriptions uses both the visuospatial and articulatory components of working memory (Brunye & Taylor, 2008). Sweller's cognitive load theory (Sweller, 2003) is concerned with the instructional consequences of a limited capacity working memory, as revealed by his research on the integration of words and pictures (Sweller & Chandler, 1994). Mayer builds on these theories and his own research to propose a model of multimedia learning (Mayer, 2001).

Pictures can also be generated as visual images to aid text comprehension. (Bergen, Lindsay, Matlock, & Narayanan, 2007). Supporting evidence for the use of visual imagery is that it takes longer to verify objects that are occluded (Horton & Rapp, 2003) such as when fog creeps into a Robert Louis Stevenson story (Stevenson, 1895/1993). Glenberg and Kaschak (2002) found direct support for visual simulation of action during comprehension of action phrases. Glenberg applied his findings to help young readers comprehend text through physical and imagined manipulation of toys (Glenberg, Gutierrez, Levin, Japuntich, & Kaschak, 2004; Glenberg, Jaworski, Rischal, & Levin, 2007). Visual simulations, however, may be used selectively for comprehension. For instance, there is evidence they are important for making predictive (Fincher-Kiefer & D'Agostino, 2004) and causally relevant inferences (Jahn, 2004).

CHAPTER 11: VISION AND ACTION

Engelkamp's (1998) book *Memory for Actions* includes an information-processing model to account for the better free recall of action phrases when the actions are performed. Reading the phrases emphasizes semantic codes and performing the actions emphasizes motor codes (Koriat & Pearlman-Avnion, 2003). The amount of activation in the motor cortex during recall increases from initially reading the action to imagining the action to performing the action (Nilsson, Nyberg, Aberg, Persson, & Roland, 2000).

Acting is a good example of a task that requires the integration of language, emotion, vision, and action. Although learning dialog could be done as strictly a verbal task, recall is better when it is integrated with the other aspects of acting (Noice & Noice, 2001, 2006). The Noices have successfully applied this active-experiencing approach to enhance the cognitive functioning of older adults (Noice, Noice, & Staines, 2004).

In addition to improving recall, action provides psychologists with evidence regarding the organization of memory that supports successful recall. Chess is a complex task in which visual thinking supports not only the selection of good moves but the ability to remember the location of pieces on the board (de Groot, 1965, 1966). Subsequent research by Chase and Simon (1973) confirmed de Groot's explanation that expert players have a perceptual advantage over weaker players because of their ability to partition the board into familiar groups (chunks).

Vision and action can also support simpler reasoning tasks such as judging whether water would pour sooner from a narrow glass or a wide glass. Most participants could not answer this question without imagining the water level as they physically tilted the glass (Schwartz & Black, 1999; Schwartz, Martin, & Nasir, 2005). Gestures are also helpful during explanations by reducing the demands on working memory (Goldin-Meadow, 2003; Richland, Zur, & Holyoak, 2007; Wagner, Nusbaum, & Goldin-Meadow, 2004). The demands are less when the gestures and verbal explanations are compatible but inconsistencies provide evidence of multiple interpretations (Garber & Goldin-Meadow, 2002; Goldin-Meadow & Wagner, 2005).

Actions and reflections on those actions play a major role in Piaget's theory (Campbell, 2001; Piaget, 1977). Although it is therefore important that manipulatives be closely linked to the concepts that students need to understand (Clements, 1999; Clements & Sarama, 2007; Richland et al., 2007; Thompson, 1994), this does not always occur in classrooms (McNeil & Jarvin, 2007; Moyer, 2002). Recent research is attempting to develop theoretical frameworks for interpreting both the strengths and limitations of manipulatives (Glenberg et al., 2007).

CHAPTER 12: VIRTUAL REALITY

Cobb and Frasier's (2005) chapter on multimedia learning in virtual reality provides a helpful introduction to this topic. The development of computer gaming is described in *The Playful World: How Technology Is Transforming Our Imagination* (Pesce, 2000). Examples of artificial reality environments include artificial surf (Grubb, 2005), the use of Sims characters in a programming course (Roth, 2006), and military simulations (Hebert, 2005). Although still in its infancy, the study of virtual worlds such as Second Life has enormous research potential (Bainbridge, 2007).

Neisser's (1976) book *Cognition and Reality* argues that the exploration of environments requires a more cyclical interplay between perception and action then is specified in traditional information processing models. Neisser was greatly influenced by the perceptual theories proposed by Gibson (1966) but argued that schematic structures are needed to guide this exploration. The role of perception and action in creating intelligence has also become increasingly important in the field of artificial intelligence (Pfeifer & Bongard, 2007). The chapters by Young (2004) and Allen, Otto, and Hoffman (2004) apply the principles of ecological psychology to educational technology. Staley's (2003) book *Computers, Visualization, and History* provides a general overview of how technology will transform our understanding of the past.

Increasing and measuring situation awareness is a common goal of many virtual reality environments. According to Klein's (1993) recognition-primed

Encapsulated Summaries

decision model, situation awareness is a key component of making good decisions. A study of electronic warfare technicians supported Klein's arguments of the importance of expertise in promoting situation awareness (Randel, Pugh, & Reed, 1996). Virtual environments are becoming increasingly useful for studying perception and action in risky environments such as making military decisions (Swartout et al., 2006), riding a bicycle (Plumert, Kearney, & Cremer, 2007), and driving a car (Pollatsek, Fisher, & Pradhan, 2006).

CHAPTER 13: SCIENCE INSTRUCTIONAL SOFTWARE

Rieber's (2003) chapter on microworlds provides an excellent overview of important work on this topic. It emphasizes the evolvement of the field, beginning with Papert's (1980) book *Mindstorms: Children, Computers, and Powerful Ideas*. One of the initial uses of microworlds was to help physics students understand Newton's laws of motion (White, 1984). The effectiveness of this approach was later demonstrated by the use of the ThinkerTools software in a sixth-grade classroom (White, 1993). A more recent version, Model-Enhanced ThinkerTools (Schwarz & White, 2005), places greater emphasis on the creation and evaluation of scientific models (also see Linn, Lee, Tinker, Husic, & Chiu, 2006; Wieman, Adams, & Perkins, 2008). Research at Carnegie Mellon University has shown that manipulating virtual objects can be as effective as manipulating real objects. There were no differences between manipulating virtual and real objects when designing science experiments (Triona & Klahr, 2003) and when assembling toy cars to maximize distance traveled (Klahr, Triona, & Williams, 2007).

A different (ecological) approach to designing microworlds specifies simple rules for describing how objects interact with each other (Resnick, 2003, 2006). The NetLogo language builds on this approach so students can create models of changes in predator–prey populations (Wilensky & Reisman, 2006). Goldstone and Son (2005) also studied complex interactions and discovered that fading from a concrete to an abstract environment facilitated the transfer of rules between different environments.

It is the field of chemistry, however, that has had the greatest amount of activity in the development and evaluation of multimedia instruction (Kozma & Russell, 2005), perhaps because of the importance of visual thinking in this field (Wu & Shah, 2004). One example is a visualization tool, eChem, that enables students to study chemical bonding by viewing three-dimensional structures of molecules (Wu, Krajcik, & Soloway, 2001). Another visualization tool, 4M:Chem, stimulates the learning of dynamic interactions by linking molecular animations with graphs, symbolic expressions, and video (Kozma, Russell, Jones, & Marx, 1996). The viewing of these dynamic interactions results in greater understanding of molecular changes than viewing a sequence of static pictures of these changes (Ardac & Akaygun, 2005).

CHAPTER 14: MATHEMATICS INSTRUCTIONAL SOFTWARE

Kaput envisioned a long-term reform of the algebra curriculum that would begin in middle school and be organized around topics such as rate and accumulation (Kaput, 1995a, 1995b, 1999). ñHe proposed that technology provided the key to making algebra accessible to all students and codeveloped SimCalc to make it happen (Roschelle, Kaput, & Stroup, 2000). Evaluation of SimCalc in mathematics classes demonstrates its effectiveness in enabling students to dynamically link representations such as simulations, graphs, and personal movement (Bowers & Nickerson, 2000; Nemirovsky, Tierney, & Wright, 1998; Nickerson, Nydam, & Bowers, 2000) that resulted in a successful scaling up to a three-week replacement unit in seventh-grade classrooms (Tater, Roschelle, Knudsen, Shechtman, Kaput, & Hopkins, 2008). Information about the SimCalc project is available at http://www.simcalc.umassd.edu.

The Animation Tutor™ also uses technology for object manipulation (Reed, 2008) and simulation in Intermediate Algebra classes. Reed (2005) provides an extensive overview of this project. There are also articles on the Average Speed (Reed & Jazo, 2002) and Task Completion (Reed, Cooke, & Jazo, 2002) modules and additional information on the Animation Tutor™ website (http://www.sci.sdsu.edu/mathtutor/). The Dimensional Thinking (Greer, Hoffman, & Reed, 2009) and Catch Up modules (Reed, Hoffman, & Phares, 2009) are part of the enclosed Animation Tutor™ DVD (Reed & Hoffman, 2009). The emphasis on modeling real-world structures is consistent with the guidelines for college algebra proposed by the Mathematical Association of America (Katz, 2007).

ANIMATE is a learning environment based on a simulation model of solving algebra word problems (Nathan, Kintsch, & Young, 1992). It provides visual feedback by simulating the equations constructed by students so they can compare the instructional simulations with their expectations based on their own mental simulations of the problem. Other applications of instructional technology offer promise for understanding geometry problems (Sedig & Sumner, 2006). Visualization software could be particularly helpful for those students who are second-language learners (Barwell, Barton, & Setati, 2007; Campbell, Adams, & Davis, 2007).

CHAPTER 15: CONCLUSIONS

Electronic learning courses are becoming increasingly widespread in high schools, universities, and businesses (Clark, 2005), as illustrated by the UC online academy for high school students (Moran, 2006). They have the *potential* to dramatically transform education and training. The importance of spatial visualization for success in mathematics and science (Lubinski & Benbow, 2006; Webb, Lubinski, & Benbow, 2007) suggests that instructional software that builds spatial skills should be particularly helpful for these courses.

An early warning about the poor showing of American students in mathematics and science was gven by the National Science Board Commission on Precollege Education in Mathematics, Science, and Technology (1983). Its more recent report *America's Pressing Challenge—Building a Stronger Foundation* (2006) shows

Encapsulated Summaries 193

that students in the United States still perform near the bottom on international assessments. One recommended solution—greater use of advanced technology—was also recommended in the Department of Commerce's (2002) *Visions 2020: Transforming Education and Training through Advanced Technologies.* The sequel, *Visions 2020.2: Student Views on Transforming Education and Training through Advanced Technologies* (2005), constructs a composite picture of students' wish lists.

Does the future offer hope for improving education through the creation of research-proven, multimedia learning environments? This is difficult to predict because as for the Amarya Indians (Lieberman, 2006; Nuñez & Sweetser, 2006), the future lies metaphorically behind us, sight unseen. There are huge challenges for applying research findings to education (Burkhardt & Schoenfeld, 2003) and for creating multimedia learning environments (*The Horizon Report*, 2006). The future does not look promising without a much greater systematic effort.

References

Ahl, V. A., Moore, C. F., & Dixon, J. A. (1992). Development of intuitive and numerical proportional reasoning. *Cognitive Development, 7*, 81–108.

Allen, B. S., Otto, R. G., & Hoffman, B. (2004). Media as lived environments: The ecological psychology of educational technology. In D. H. Jonassen (Ed.), *Handbook of research for educational communications and technology* (2nd ed., pp. 215–241). Mahwah, NJ: Lawrence Erlbaum Associates.

Anglin, G. J., Vaez, H., & Cunningham, K. L. (2003). Visual representations and learning: The role of static and animated graphics. In D. Jonassen (Ed.), *Handbook of research on educational communications and technology* (2nd ed., pp. 865–916). Mahwah, NJ: Lawrence Erlbaum Associates.

Ardac, D., & Akaygun, S. (2005). Using static and dynamic visuals to represent chemical change at the molecular level. *International Journal of Science Education, 27*, 1269–1298.

Baddeley, A. D. (1992). Is working memory working? The fifteenth Bartlett lecture. *Quarterly Journal of Experimental Psychology, 44A*, 1–31.

Baddeley, A. D. (2000). The episodic buffer: A new component of working memory? *Trends in Cognitive Sciences, 4*, 417–423.

Baddeley, A. D. (2001). Is working memory still working? *American Psychologist, 56*, 851–864.

Baddeley, A. D., & Hitch, G. (1974). Working memory. In G. H. Bower (Ed.), *The psychology of learning and motivation* (Vol. 8, pp. 17–90). Orlando, FL: Academic Press.

Baker, E., Dickieson, J., Wulfeck, W. & O'Neil, H. F. (Eds.). (2008). Assessment of problem solving using simulations. New York: Lawrence Erlbaum Associates.

Bainbridge, W. S. (2007). The scientific research potential of virtual worlds. *Science, 317*, 472–476.

Barsalou, L. W. (1999). Perceptual symbol systems. *Behavioral & Brain Sciences, 22*, 577–660.

Barsalou, L. W., Simmons, W. K., Barbey, A. K., & Wilson, C. D. (2003). Grounding conceptual knowledge in modality-specific systems. *TRENDS in Cognitive Sciences, 7*, 84–91.

Barwell, R., Barton, B., & Setati, M. (2007). Multilingual issues in mathematics education: Introduction. *Educational Studies in Mathematics, 64*, 113–119.

Bentall, R. P. (1990). The illusion of reality: A review and integration of psychological research on hallucinations. *Psychological Bulletin, 107*, 82–95.

Bergen, B. K., Lindsay, S., Matlock, T., & Narayanan, S. (2007). Spatial and linguistic aspects of visual imagery in sentence comprehension. *Cognitive Science, 31*, 733–764.

Booth, J. L., & Siegler, R. S. (2006). Developmental and individual differences in pure numerical estimation. *Developmental Psychology, 41*, 189–201.

Bowers, J. S., & Nickerson, S. N. (2000). Students' changing views of rates and graphs when working with a simulation microworld. *Focus on Learning Problems in Mathematics, 22*, 10–25.

Brandimonte, M. A., & Gerbino, W. (1993). Mental image reversal and verbal recoding: When ducks become rabbits. *Memory & Cognition, 21*, 23–33.

Brooks, L. R. (1968). Spatial and verbal components of the act of recall. *Canadian Journal of Psychology, 22*, 349–368.

Brown, D. (2003). *The Da Vinci code: A novel*. New York: Doubleday.
Brunye, T. T., & Taylor, H. A. (2008). Working memory in developing and applying mental models from spatial descriptions. *Journal of Memory and Language, 58,* 701–729.
Burkhardt, H., & Schoenfeld, A. H. (2003). Improving educational research: Toward a more useful, more influential, and better-funded enterprise. *Educational Researcher, 32,* 3–14.
Calvo, M. G., & Lang, P. J. (2004). Gaze patterns when looking at emotional pictures: Motivationally biased attention. *Motivation and Emotion, 28,* 221–243.
Campbell, A. E., Adams, V. M., & Davis, G. E. (2007). Cognitive demands and second-language learners: A framework for analyzing mathematics instructional contexts. *Mathematical Thinking and Learning, 9,* 3–30.
Campbell, R. L. (Ed.). (2001). *Studies in reflecting abstraction.* Hove, England: Psychology Press.
Capote, T. (1965). *In cold blood.* New York: Random House.
Carlson, M., Jacobs, S., Coe, E., Larsen, S., & Hsu, E. (2002). Applying covariational reasoning while modeling dynamic events: A framework and a study. *Journal for Research in Mathematics Education, 33,* 352–378.
Carroll, J. M., Thomas, J. C., & Malhotra, A. (1980). Presentation and representation in design problem solving. *British Journal of Psychology, 71,* 143–153.
Casasanto, D., & Boroditsky, L. (2008). Time in mind: Using space to think about time. *Cognition, 106,* 579–593.
Cemen, P. B. (1993). Adding and subtracting integers on the number line. *The Arithmetic Teacher, 40,* 388–389.
Chambers, D., & Reisberg, D. (1985). Can mental images be ambiguous? *Journal of Experimental Psychology: Human Perception and Performance, 11,* 317–328.
Chambers, D., & Reisberg, D. (1992). What an image depicts depends on what an image means. *Cognitive Psychology, 24,* 145–174.
Chase, W. G., & Simon, H. A. (1973). Perception in chess. *Cognitive Psychology, 4,* 55–81.
Chatterjee, A. (2001). Language and space: Some interactions. *TRENDS in Cognitive Sciences, 5,* 55–61.
Chen, W., Kato, T., Shu, X. H., Ogawa, S., Tank, D. W., & Ugurbil, K. (1998). Human primary visual cortex and lateral geniculate nucleus activation during visual imagery. *Neuroreport, 9,* 3669–3674.
Clark, R. C. (2005). Multimedia learning in e-courses. In R. E. Mayer (Ed.), *The Cambridge handbook of multimedia learning* (pp. 589–616). Cambridge, England: Cambridge University Press.
Clements, D. H. (1999). "Concrete" manipulatives, concrete ideas. *Contemporary Issues in Early Childhood, 1,* 45–60.
Clements, D. H., & Sarama, J. (2007). Effects of a preschool mathematics curriculum: Summative research on the Building Blocks project. *Journal for Research in Mathematics Education, 38,* 136–163.
Cobb, S., & Fraser, D. S. (2005). Multimedia learning in virtual reality. In R. E. Mayer (Ed.), *The Cambridge handbook of multimedia learning* (pp. 525–547). Cambridge, England: Cambridge University Press.
Cohen, J. E. (2000). Population growth and earth's human carrying capacity. In A.R. Chapman, R.L. Peterson, & B. Smith-Moran (Eds.). *Consumption, population, and sustainability: Perspectives from science and religion.* Washington, DC: Island Press.
Cole, K. C. (1998). *The universe and the teacup: The mathematics of truth and beauty.* Orlando, FL: Harcourt Brace & Company.

References

Collins, A. M., & Loftus, E. F. (1975). A spreading activation theory of semantic processing. *Psychological Review, 82*, 407–428.
Conroy, P. (1986). *The prince of tides*. Boston: Houghton Mifflin.
Cornell, E. (1974). Infants' discrimination of photographs of faces following redundant presentation. *Journal of Experimental Child Psychology, 18*, 98–106.
Davis, R. (1998). What is intelligence? Why? *AI Magazine, 19*, 91–110.
Davis, R., Shrobe, H., & Szolovits, P. (1993). What is a knowledge representation? *AI Magazine, 14*, 17–33.
Dawson, M., Soulieres, I., Gernsbacher, M. A., & Mottron, L. (2007). The level and nature of autistic intelligence. *Psychological Science, 18*, 657–662.
de Groot, A. D. (1965). *Thought and choice in chess*. New York: Basic Books.
de Groot, A. D. (1966). Perception and memory versus thought: Some old ideas and recent findings. In B. Kleinmuntz (Ed.), *Problem solving: Research, method, and theory*. New York: Wiley.
Dehaene, S. (1997). *The number sense: How the mind creates mathematics*. New York: Oxford University Press.
Dehaene, S., Molko, N., Cohen, L., & Wilson, A. J. (2004). Arithmetic and the brain. *Current Opinion in Neurobiology, 14*, 218–224.
Dehaene, S., Spelke, E., Pinel, P., Stanescu, R., & Tsivkin, S. (1999). Sources of mathematical thinking: Behavioral and brain-imaging evidence. *Science, 284*, 970–974.
DeLoache, J. S. (2004). Becoming symbol minded. *TRENDS in Cognitive Sciences, 8*, 66–70.
Diezmann, C., & English, L. D. (2001). Promoting the use of diagrams as tools for thinking. In A. A. Cuoco & F. R. Curcio (Eds.), *The roles of representation in school mathematics* (pp. 77–89). Reston, VA: National Council of Teachers of Mathematics.
diSessa, A. A. (2004). Metarepresentation: Native competence and targets for instruction. *Cognition and Instruction, 22*, 293–331.
Dixon, J. A., & Moore, C. F. (1996). The developmental role of intuitive principles in choosing mathematical strategies. *Developmental Psychology, 32*, 241–253.
Doll, R. (1955). Etiology of lung cancer. *Advances in Cancer Research, 3*, 1–50.
Dove, G. (2009). Beyond perceptual symbols: A call for representational pluralism. *Cognition, 110*, 412–431.
Engelkamp, J. (1998). *Memory for actions*. Hove, England: Psychology Press.
Fagan, J. F. (1973). Infants' delayed recognition memory and forgetting. *Journal of Experimental Child Psychology, 16*, 424–450.
Fantz, R. L. (1961). The origin of form perception. *Scientific American, 204*, 66–72.
Feigenson, L., Carey, S., & Hauser, M. (2002). The representations underlying infants' choice of more: Object files versus analogy magnitudes. *Psychological Science, 13*, 150–156.
Feigenson, L., Dehaene, S., & Spelke, E. (2004). Core systems of number. *TRENDS in Cognitive Sciences, 8*, 307–314.
Fincher-Kiefer, R., & D'Agostino, P. R. (2004). The role of visuospatial resouces in generating predictive and bridging inferences. *Discourse Processes, 37*, 205–224.
Finke, R. A. (1990). *Creative imagery: Discoveries and inventions in visualization*. Mahwah, NJ: Lawrence Erlbaum Associates.
Finke, R. A., Ward, T. B., & Smith, S. M. (1992). *Creative cognition: Theory, research, and applications*. Cambridge, MA: The MIT Press.
Friel, S. N., Curcio, F. R., & Bright, G. W. (2001). Making sense of graphs: Critical factors influencing comprehension and instructional implications. *Journal for Research in Mathematics Education, 32*, 124–158.

Gallistel, C. R., & Gelman, R. (2000). Non-verbal numerical cognition: From reals to integers. *TRENDS in Cognitive Sciences, 4,* 59–65.
Garber, P., & Goldin-Meadow, S. (2002). Gesture offers insight into problem-solving in adults and children. *Cognitive Science, 26,* 817–831.
Gelman, R., & Gallistel, C. R. (2004). Language and the origin of numerical concepts. *Science, 306,* 441–443.
Gibbons, K. (2006). *The life all around me by Ellen Foster.* Orlando, FL: Harcourt.
Gibbs, R. W. (2006). *Embodiment and cognitive science.* New York: Cambridge University Press.
Gibson, J. J. (1966). *The senses considered as perceptual systems.* Boston: Houghton Mifflin.
Gleitman, L., & Papafragou, A. (2005). Language and thought. In K. J. Holyoak & R. G. Morrison (Eds.), *The Cambridge handbook of thinking and reasoning* (pp. 633–661). Cambridge, England: Cambridge University Press.
Glenberg, A. M., Gutierrez, T., Levin, J. R., Japuntich, S., & Kaschak, M. P. (2004). Activity and imagined activity can enhance young children's reading comprehension. *Journal of Educational Psychology, 96,* 424–436.
Glenberg, A. M., Jaworski, B., Rischal, M., & Levin, J. R. (2007). What brains are for: Action, meaning, and reading comprehension. In D. McNamara (Ed.), *Reading comprehension strategies: Theories, interventions, and technologies* (pp. 221–240). Mahwah, NJ: Lawrence Erlbaum Associates.
Glenberg, A. M., & Kaschak, M. P. (2002). Grounding language in action. *Psychonomic Bulletin & Review, 9,* 558–565.
Goldin-Meadow, S. (2003). *Hearing gesture: How our hands help us think.* Cambridge, MA: Harvard University Press.
Goldin-Meadow, S., & Wagner, S. M. (2005). How our hands help us learn. *TRENDS in Cognitive Sciences, 9,* 234–241.
Goldstone, R. L., & Son, J. Y. (2005). The transfer of scientific principles using concrete and idealized simulations. *The Journal of the Learning Sciences, 14,* 69–110.
Gorodetsky, M., & Fisher, K. M. (1996). Generating connections and learning in biology. In K. M. Fisher & M. R. Kibby (Eds.), *Knowledge acquisition, organization, and use in biology* (pp. 135–154). Berlin: Springer.
Grandin, T., & Johnson, C. (2005). *Animals in translation.* Orlando, FL: Harcourt.
Grant, E. R., & Spivey, M. J. (2003). Eye movements and problem solving: Guiding attention guides thought. *Psychological Science, 14,* 462–466.
Greer, B. (1993). The mathematical modeling perspective on wor(l)d problems. *Journal of Mathematical Behavior, 12,* 239–250.
Greer, B., Hoffman, B., & Reed, S. K. (2009). Animation Tutor: Dimensional thinking [Computer software]. San Diego, CA: San Diego State University.
Grubb, S. (2005, July 21). Making waves. *San Diego Union-Tribune,* p. 25.
Hajela, D. (2005, October 19). The covers that captured attention. *San Diego Union-Tribune.*
Harp, S. F., & Mayer, R. E. (1997). The role of interest in learning from scientific text and illustrations: On the distinction between emotional interest and cognitive interest. *Journal of Educational Psychology, 89,* 92–102.
Hauser, M. D. (2000). What do animals think about numbers? *American Scientist, 88,* 144–151.
Hebert, J. (2005, November 6). Band of brothers. *San Diego Union-Tribune,* p. F1.
Hegarty, M. (1992). Mental animation: Inferring motion from static displays of mechanical systems. *Journal of Experimental Psychology: Learning, Memory and Cognition, 18,* 1084–1102.
Hegarty, M. (2004). Mechanical reasoning by mental simulation. *TRENDS in Cognitive Sciences, 8,* 280–285.

References

Hegarty, M., & Kozhevnikov, M. (1999). Types of visual-spatial representations and mathematical problem solving. *Journal of Educational Psychology, 91*, 684–689.

Hegarty, M., Kriz, S., & Cate, C. (2003). The roles of mental animations and external animations in understanding mechanical systems. *Cognition and Instruction, 21*, 325–360.

Heidbreder, E. (1961). *Seven psychologies.* New York: Appleton-Century-Crofts.

Hesslow, G. (2002). Conscious thought as simulation of behavior and perception. *TRENDS in Cognitive Sciences, 6*, 242–247.

Hidi, S., & Renninger, K. A. (2006). The four-phase model of interest development. *Educational Psychologist, 41*, 111–127.

Hirshberg, J. (1998). *The creative priority.* New York: Harper Collins.

Höffler, T. N., & Leutner, D. (2007). Instructional animation versus static pictures: A meta-analysis. *Learning and Instruction, 17*, 722–738.

Hoffman, B. (2000). *The mystery of the mission museum.* Instructional software. San Diego, CA: San Diego State University.

Hogan, T. P., & Brezinski, K. L. (2003). Quantitative estimation: One, two, or three abilities? *Mathematical Thinking and Learning, 5*, 259–280.

Holley, C. D., & Dansereau, D. F. (1984). Networking: The technique and empirical evidence. In C. D. Holley & D. F. Dansereau (Eds.), *Spatial learning strategies.* Orlando, FL: Academic Press.

Horton, W. S., & Rapp, D. N. (2003). Out of sight, out of mind: Occlusion and the accessibility of information in narrative comprehension. *Psychonomic Bulletin & Review, 10*, 104–110.

Hurley, S. M., & Novick, L. R. (2006). Context and structure: The nature of students' knowledge about three spatial diagram representations. *Thinking & Reasoning, 12*, 281–308.

Hyman, I. E., & Pentland, J. (1996). The role of mental imagery in the creation of false childhood memories. *Journal of Memory and Language, 35*, 101–117.

Inhelder, B., & Piaget, J. (1958). *The growth of logical thinking from childhood to adolescence.* New York: Basic Books.

Jahn, G. (2004). Three turtles in danger: Spontaneous construction of causally relevant spatial situation models. *Journal of Experimental Psychology: Learning, Memory and Cognition, 30*, 969–987.

Johnson, L. F. & Smith, R. S. (2006). 2006 Horizon Report. Austin, TX: The New Media Consortium.

Johnson, M. (1987). *The body in the mind: The bodily basis of meaning, imagination, and reasoning.* Chicago: The University of Chicago Press.

Johnson, M. K., & Raye, C. L. (1981). Reality monitoring. *Psychological Review, 88*, 67–85.

Johnson, M. K., Raye, C. L., Wang, A. Y., & Taylor, T. T. (1979). Fact and fantasy: The roles of accuracy and variability in confusing imaginations with perceptual experiences. *Journal of Experimental Psychology: Human Learning and Memory, 5*, 229–240.

Joram, E., Gabriele, A. J., Bertheau, M., Gelman, R., & Subrahmanyam, K. (2005). Children's use of the reference point strategy for measurement estimation. *Journal for Research in Mathematics Education, 36*, 4–23.

Joram, E., Subrahmanyam, K., & Gelman, R. (1998). Measurement estimation: Learning to map the route from number to quantity and back. *Review of Educational Research, 68*, 413–449.

Kaput, J. J. (1995a). Long-term algebra reform: Democratizing access to big ideas. In C. B. Lacampagne, W. Blair, & J. Kaput (Eds.), *The algebra initiative colloquium* (Vol. 1, pp. 33–49). Washington, DC: U.S. Department of Education.

Kaput, J. J. (1995b). *A research base supporting long-term algebra reform*. Paper presented at the Seventeenth Annual Meeting of the North American Chapter of the International Group for the Psychology of Mathematics Education, Columbus, Ohio.

Kaput, J. J. (1999). Teaching and learning a new algebra. In E. Fennema & T. A. Romberg, (Eds.), *Mathematics classrooms that promote understanding* (pp. 133–155). Mahwah, NJ: Lawrence Erlbaum Associates.

Katz, V. J. (Ed.). (2007). *Algebra: Gateway to a technological future*. Washington, DC: Mathematical Association of America.

Kiesel, A., & Vierck, E. (2009). SNARC-like congruency based on number magnitude and response duration. *Journal of Experimental Psychology: Learning, Memory and Cognition, 35*, 275–279.

Klahr, D., Triona, L. M., & Williams, C. (2007). Hands on what? The relative effectiveness of physical vs. virtual materials in an engineering design project by middle school children. *Journal of Research in Science Teaching, 44*, 183–203.

Klein, G. A. (1993). A recognition-primed decision (RPD) model of rapid decision making. In G. A. Klein, R. Orasanu, R. Calderwood, & C. E. Zsambok (Eds.), *Decision making in action: Models and methods*. Norwood, NJ: Ablex Publishing Corporation.

Knoblich, G., Ohlsson, S., Haider, H., & Rhenius, D. (1999). Constraint relaxation and chunk decomposition in insight problem solving. *Journal of Experimental Psychology: Learning, Memory and Cognition, 25*, 1534–1555.

Kohler, W. (1925). *The mentality of apes*. New York: Harcourt.

Koriat, A., & Pearlman-Avnion, S. (2003). Memory organization of action events and its relationship to memory performance. *Journal of Experimental Psychology: General, 132*, 435–454.

Kosslyn, S. M., Ball, T. M., & Reiser, B. J. (1978). Visual images preserve metric spatial information: Evidence from studies of visual scanning. *Journal of Experimental Psychology: Human Perception and Performance, 4*, 47–60.

Kosslyn, S. M., & Pomerantz, J. R. (1977). Imagery, propositions, and the form of internal representations. *Cognitive Psychology, 9*, 52–76.

Kosslyn, S. M., Thompson, W. L., & Ganis, G. (2006). *The case for mental imagery*. New York: Oxford University Press.

Kozhevnikov, M., Hegarty, M., & Mayer, R. E. (2002). Revising the visualizer-verbalizer dimension: Evidence for two types of visualizers. *Cognition and Instruction, 20*, 47–77.

Kozma, R., & Russell, J. (2005). Multimedia learning of chemistry. In R. E. Mayer (Ed.), *The Cambridge handbook of multimedia learning* (pp. 409–428). Cambridge, England: Cambridge University Press.

Kozma, R., Russell, J., Jones, T., & Marx, N. (1996). The use of multiple, linked representations to facilitate understanding of chemistry. In S. Vosniadou, E. De Corte, R. Glaser, & H. Mandl (Eds.), *International perspectives on the design of technology-supported learning environments* (pp. 41–60). Mahwah, NJ: Lawrence Erlbaum Associates.

Lafleur, L. J. (1960). Translation of R. Descartes' discourse on method and meditations. New York: The Liberal Arts Press.

Lakoff, G. (1987). *Women, fire, and dangerous things: What categories reveal about mind*. Chicago: University of Chicago Press.

Lakoff, G., & Johnson, M. (1980). *Metaphors we live by*. Chicago: University of Chicago Press.

Lakoff, G., & Nuñez, R. E. (1997). The metaphorical structure of mathematics: Sketching out cognitive foundations for a mind-based mathematics. In L. D. English (Ed.), *Mathematical reasoning: Analogies, metaphors, and images* (pp. 21–89). Mahwah, NJ: Lawrence Erlbaum Associates.

Lakoff, G., & Nuñez, R. E. (2000). *Where mathematics comes from: How the embodied mind brings mathematics into being*. New York: Basic Books.

Levin, J. R. (1989). A transfer-appropriate-processing perspective of pictures in prose. In H. Mandl & J. R. Levin (Eds.), *Knowledge acquisition from text and pictures*. Amsterdam: Elsevier.

Lieberman, B. (2006, June 22). Time flies—ahead of the Aymara. *San Diego Union-Tribune*, p. B1.

Lindsay, D. S., & Read, D. D. (1994). Psychotherapy and memories of childhood sexual abuse: A cognitive perspective. *Applied Cognitive Psychology, 8*, 281–338.

Linn, M. C., Lee, H., Tinker, R., Husic, F., & Chiu, J. L. (2006). Teaching and assessing knowledge integration in science. *Science, 313*, 1049–1050.

Lobato, J. (1996). Transfer reconceived: How "sameness" is produced in mathematical activity. Unpublished doctoral dissertation, University of California, Berkeley, CA.

Lobato, J. (2008). When students don't apply the knowledge you think they have, rethink your assumptions about transfer. In M. Carlson & C. Rasmussen (Eds.), *Making the connection: Research and teaching in undergraduate mathematics* (pp. 289–304). Washington, DC: Mathematical Association of America.

Loftus, E. F. (1993). The reality of repressed memories. *American Psychologist, 48*, 518–537.

Lorayne, H., & Lucas, J. (1974). *The memory book*. New York: Ballantine.

Lubinski, D., & Benbow, C. P. (2006). Study of mathematically precocious youth after 35 years: Uncovering antecedents for the development of math-science expertise. *Perspectives on Psychological Science, 1*, 316–345.

Mandler, J. M. (2004). *The foundations of mind: Origins of conceptual thought*. Oxford, England: Oxford University Press.

Mandler, J. M., & McDonough, L. (1996). Drinking and driving don't mix: Inductive generalization in infancy. *Cognition, 59*, 307–355.

Markman, A. B. (1999). *Knowledge representation*. Mahwah, NJ: Lawrence Erlbaum Associates.

Matlock, T. (2004). Fictive motion as cognitive simulation. *Memory & Cognition, 32*, 1389–1400.

Mayer, R. E. (2001). *Multimedia learning*. Cambridge: Cambridge University Press.

McNeil, N. M., & Jarvin, L. (2007). When theories don't add up: Disentangling the manipulatives debate. *Theory Into Practice, 46*, 309–316.

Metcalf, J. (1986). Premonitions of insight predict impending error. *Journal of Experimental Psychology: Learning, Memory and Cognition, 12*, 623–634.

Miller, A. I. (1984). *Imagery in scientific thought*. Boston: Birkhauser.

Mitchell, D. B., & Richman, C. L. (1980). Confirmed reservations: Mental travel. *Journal of Experimental Psychology: Human Perception and Performance, 6*, 58–66.

Moran, C. (2006, July 5). UC plans to launch an online academy. *San Diego Union-Tribune*, p. B1.

Moreno, R. & Meyer, R. (2007). Interactive multimodel learning environments. *Educational Psychology Review, 19*, 309–326.

Morris, P. E., Jones, S., & Hampson, P. (1978). An imagery mnemonic for the learning of people's names. *British Journal of Psychology, 69*, 335–336.

Moyer, P. S. (2002). Are we having fun yet? How teachers use manipulatives to teach mathematics. *Educational Studies in Mathematics, 47*, 175–197.

Moyer, R. S., & Landauer, T. K. (1967). Time required for judgments of numerical inequality. *Nature, 215*, 1519–1520.

Nathan, M. J., Kintsch, W., & Young, E. (1992). A theory of algebra-word-problem comprehension and its implications for the design of learning environments. *Cognition and Instruction, 9*, 329–389.

National Science Board Commission on Precollege Education in Mathematics, Science, and Technology. (1983). Educating Americans for the 21st Century. Washington, DC: National Science Foundation.

National Science Board Commission on Precollege Education in Mathematics, Science, and Technology. (2006). America's Pressing Challenge—Building a Stronger Foundation. Arlington, VA: National Science Foundation.

Neisser, U. (1967). *Cognitive psychology*. New York: Appleton-Century-Crofts.

Neisser, U. (1976). *Cognition and reality*. San Francisco: W. H. Freeman and Company.

Nemirovsky, R., Tierney, C., & Wright, T. (1998). Body motion and graphing. *Cognition and Instruction, 16*, 119–172.

Newell, A., & Simon, H. A. (1972). *Human problem solving*. Englewood Cliffs, NJ: Prentice-Hall.

Nickerson, S. N., Nydam, C., & Bowers, J. S. (2000). Linking algebraic concepts and contexts: Every picture tells a story. *Mathematics Teaching in Middle School, 6*, 92–98.

Nielsen, G. D., & Smith, E. E. (1973). Imaginal and verbal representations in short-term recognition of visual faces. *Journal of Experimental Psychology, 101*, 375–378.

Nilsson, L.-G., Nyberg, L., Klingberg, T., Aberg, C., Persson, J., & Roland, P. E. (2000). Activity in motor areas while remembering action events. *Neuroreport, 11*, 2199–2201.

Noice, H., & Noice, T. (2001). Learning dialogue with and without movement. *Memory & Cognition, 29*, 820–827.

Noice, H., & Noice, T. (2006). What studies of actors and acting can tell us about memory and cognitive functioning. *Current Directions in Psychological Science, 15*, 14–18.

Noice, H., Noice, T., & Staines, G. (2004). A short-term intervention to enhance cognitive and affective functioning in older adults. *Journal of Aging and Health, 16*, 1–24.

Norretranders, T. (1998). *The user illusion*. New York: Penguin Putnam Inc.

Novick, L. R. (2006). Understanding spatial diagram structure: An analysis of hierarchies, matrices, and networks. *The Quarterly Journal of Experimental Psychology, 59*, 1826–1856.

Novick, L. R., & Hurley, S. M. (2001). To matrix, network, or hierarchy: That is the question. *Cognitive Psychology, 42*, 158–216.

Novick, L. R., Hurley, S. M., & Francis, M. (1999). Evidence for abstract, schematic knowledge of three spatial diagram representations. *Memory & Cognition, 27*, 288–308.

Nuñez, R. E., Edwards, L. D., & Matos, J. F. (1999). Embodied cognition as grounding for situatedness and context in mathematics education. *Educational Studies in Mathematics, 39*, 45–65.

Nuñez, R. E., & Sweetser, E. (2006). With the future behind them: Convergent evidence from Aymara language and gesture in the crosslinguistic comparison of spatial construals of time. *Cognitive Science, 30*, 401–450.

O'Donnell, A. M., Dansereau, D. F., & Hall, R. F. (2002). Knowledge maps as scaffolds for cognitive processing. *Educational Psychology Review, 14*, 71–86.

Paivio, A. (1969). Mental imagery in associative learning and memory. *Psychological Review, 76*, 241–263.

Paivio, A. (1986). *Mental representations: A dual coding approach*. New York: Oxford University Press.

Palmer, S. E. (1978). Fundamental aspects of cognitive representation. In E. Rosch & B. B. Lloyd (Eds.), *Cognition and categorization*. Mahwah, NJ: Lawrence Erlbaum Associates.

Pani, J. R. (1996). Mental imagery as the adaptionist views it. *Consciousness and Cognition, 5*, 288–326.

Papert, S. (1980). *Mindstorms: Children, computers, and powerful ideas*. New York: Basic Books.

Park, O., & Hopkins, R. (1993). Instructional conditions for using dynamic visual displays: A review. *Instructional Science, 22*, 1–24.

Pesce, M. (2000). *The playful world: How technology is transforming our imagination*. New York: Ballantine Books.

Pfeifer, R., & Bongard, J. (2007). *How the body shapes the way we think: A new view of intelligence*. Cambridge, MA: The MIT Press.

Philipp, R. A., Martin, W. O., & Richgels, G. W. (1993). Curricular implications of graphical representations of functions. In T. A. Romberg, E. Fennema, & T. P. Carpenter (Eds.), *Integrating research on graphical representations of functions*. Mahwah, NJ: Lawrence Erlbaum Associates.

Piaget, J. (1977). *Recherches sur l'abstraction réfléchissante*. Paris: Presses Universitaires de France.

Pinker, S. (1994). *The language instinct*. New York: William Morrow.

Plass, J. L., Homer, B. D., & Hayward, E. O. (2009). Design factors for educationally effective animations and simulations. *Journal of Computing in Higher Education, 21*, 31–61.

Plumert, J. M., Kearney, J. K., & Cremer, J. F. (2007). Children's road crossing: A window into perceptual-motor development. *Current Directions in Psychological Science, 16*, 255–263.

Pollatsek, A., Fisher, D. L., & Pradhan, A. (2006). Identifying and remedying failures of selective attention in younger drivers. *Current Directions in Psychological Science, 15*, 255–259.

Pylyshyn, Z. (1973). What the mind's eye tells the mind's brain: A critique of mental imagery. *Psychological Bulletin, 80*, 1–24.

Pylyshyn, Z. (1981). The imagery debate: Analogue media versus tacit knowledge. *Psychological Review, 88*, 16–45.

Pylyshyn, Z. (2003). Return of the mental image: Are there really pictures in the brain? *TRENDS in Cognitive Sciences, 7*, 113–118.

Randel, J. M., Pugh, H. L., & Reed, S. K. (1996). Differences in expert and novice situation awareness in naturalistic decision making. *International Journal of Human-Computer Studies, 45*, 579–597.

Reed, S. K. (1972). Pattern recognition and categorization. *Cognitive Psychology, 3*, 382–407.

Reed, S. K. (1974). Structural descriptions and the limitations of visual imagery. *Memory & Cognition, 2*, 329–336.

Reed, S. K. (1982). *Cognition: Theory and applications*. Monterey, CA: Brooks/Cole.

Reed, S. K. (1999). *Word problems: Research and curriculum reform*. Mahwah, NJ: Lawrence Erlbaum Associates.

Reed, S. K. (2005). From research to practice and back: The Animation Tutor™ project. *Educational Psychology Review, 17*, 55–82.

Reed, S. K. (2006). *Cognition: Theory and applications* (7th ed.). Belmont, CA: Wadsworth/Thomson Learning.

Reed, S. K. (2008). Manipulating multimedia materials. In R. Zheng (Ed.), *Cognitive effects of multimedia learning* (pp. 51–66). New York: IGI Global.

Reed, S. K., Cooke, J., & Jazo, L. (2002). Building complex solutions from simple solutions in the Animation Tutor: Task completion. *Mathematical Thinking and Learning, 4*, 315–336.

Reed, S. K., & Evans, A. C. (1987). Learning functional relations: A theoretical and instructional analysis. *Journal of Experimental Psychology: General, 116*, 106–118.
Reed, S. K., & Friedman, M. P. (1973). Perceptual vs. conceptual categorization. *Memory & Cognition, 1*, 157–163.
Reed, S. K., Hock, H., & Lockhead, G. R. (1983). Tacit knowledge and the effect of pattern configuration on mental scanning. *Memory & Cognition, 3*, 569–575.
Reed, S. K., & Hoffman, B. (2004). Use of temporal and spatial information in estimating event completion time. *Memory & Cognition, 32*, 271–282.
Reed, S. K., & Hoffman, B. (2009). Animation Tutor [computer software]. San Diego, CA: San Diego State University.
Reed, S. K., Hoffman, B., & Phares, S. (2009). Animation Tutor: Catch up [Computer software]. San Diego, CA: San Diego State University.
Reed, S. K., Hoffman, B., & Short, D. (2009). Animation Tutor: Population growth [Computer software]. San Diego, CA: San Diego State University.
Reed, S. K., & Jazo, L. (2002). Using multiple representations to improve conceptions of average speed. *Journal of Educational Computing Research, 27*, 147–166.
Resnick, M. (2003). Thinking like a tree (and other forms of ecological thinking). *International Journal of Computers for Mathematical Learning, 8*, 43–62.
Resnick, M. (2006). Computer as paintbrush: Technology, play, and the creative society. In D. Singer, R. Golikoff, & K. Hirsh-Pasek (Eds.), *Play = learning: How play motivates and enhances children's cognitive and social-emotional growth*. New York: Oxford University Press.
Richland, L. E., Zur, O., & Holyoak, K. J. (2007). Cognitive supports for analogies in the mathematics classroom. *Science, 316*, 1128–1129.
Rieber, L. P. (1990). Animation in computer-based instruction. *Educational Technology, Research and Development, 38*, 77–86.
Rieber, L. P. (2003). Microworlds. In D. Jonassen (Ed.), *Handbook of research on educational communications and technology* (2nd ed., pp. 583–603). Mahwah, NJ: Erlbaum.
Rips, L. J., Shoben, E. J., & Smith, E. E. (1973). Semantic distance and the verification of semantic relations. *Journal of Verbal Learning and Verbal Behavior, 12*, 1–20.
Rizzo, A., Reger, G., Gahm, G., Difede, J., & Rothbaum, B. O. (2009). Virtual reality exposure therapy for combat-related PTSD. In P. J. Shiromani, T. M. Keane, & J. E. LeDoux (Eds.), *Post-traumatic stress disorder: Basic science and clinical practice*. Totowa, NJ: Humana Press.
Roschelle, J., Kaput, J. J., & Stroup, W. (2000). SimCalc: Accelerating students' engagement with the mathematics of change. In M. J. Jacobson & R. B. Kozma (Eds.), *Innovations in science and mathematics education: Advanced designs for technologies of learning* (pp. 48–75). Mahwah, NJ: Erlbaum.
Roth, M. (2006, May 29). Sims characters could attract more women to programming. *San Diego Union-Tribune*, p. E2.
Rubin, D. C. (2007). The basic-systems model of episodic memory. *Perspectives on Psychological Science, 1*, 277–311.
Saariluoma, P. (1992). Visuospatial and articulatory interference in chess players' information intake. *Applied Cognitive Psychology, 6*, 77–89.
Sack, A. T., van de Ven, V. G., Etschenberg, S., & Linden, D. E. (2005). Enhanced vividness of mental imagery as a trait marker of schizophrenia. *Schizophrenia Bulletin, 31*, 1–8.
Schooler, J. W., & Engstler-Schooler, T. Y. (1990). Verbal overshadowing of visual memories: Some things are better left unsaid. *Cognitive Psychology, 17*, 36–71.
Schwartz, D. L., & Black, J. B. (1996). Shuttling between depictive models and abstract rules: Induction and fallback. *Cognitive Science, 20*, 457–497.

Schwartz, D. L., & Black, T. (1999). Inferences through imagined actions: Knowing by simulated doing. *Journal of Experimental Psychology: Learning, Memory and Cognition, 25*, 116–136.

Schwartz, D. L., Blair, K. P., Biswas, G., Leeawong, K., & Davis, J. (2008). Animations of thought: Interactivity in the teachable agent paradigm. In R. Lowe & W. Schnotz (Eds.), *Learning with animation: Research and implications for design*. Cambridge, England: Cambridge University Press.

Schwartz, D. L., & Heiser, J. (2006). Spatial representations and imagery in learning. In *The Cambridge handbook of the learning sciences* (pp. 283–298). Cambridge, England: Cambridge University Press.

Schwartz, D. L., Martin, T., & Nasir, N. (2005). Designs for knowledge evolution: Towards a prescriptive theory of integrating first- and second-hand knowledge. In P. Gardenfors & P. Johansson (Eds.), *Cognition, education, and communication technology* (pp. 21–54). Mahwah, NJ: Lawrence Erlbaum Associates.

Schwarz, C. V., & White, B. Y. (2005). Metamodeling knowledge: Developing students' understanding of scientific modeling. *Cognition and Instruction, 23*, 165–205.

Sedig, K., & Sumner, M. (2006). Characterizing interaction with visual mathematical representations. *International Journal of Computers for Mathematical Learning, 11*, 1–55.

Shepard, R. N. (1957). Stimulus and response generalization: A stochastic model representing generalization to distance in psychological space. *Psychometrica, 22*, 325–345.

Shepard, R. N. (1988). The imagination of the scientist. In K. Egan & D. Nadaner (Eds.), *Imagination and education* (pp. 153–185). New York: Teacher's College Press.

Shepard, R. N., & Metzler, J. (1971). Mental rotation of three-dimensional objects. *Science, 171*, 701–703.

Siegler, R. S., & Opfer, J. E. (2003). The development of numerical estimation: Evidence of multiple representations of numerical quantity. *Psychological Science, 14*, 237–243.

Siegler, R. S., & Ramani, G. (2006). Early development of estimation skills. *Observer, 19*, 34–44.

Simon, H. A., & Reed, S. K. (1976). Modeling strategy shifts in a problem-solving task. *Cognitive Psychology, 8*, 86–97.

Singer, J. L. (Ed.). (1990). *Repression and dissociation: Implications for personality theory, psychopathology, and health*. Chicago: University of Chicago Press.

Smith, E. E., & Nielsen, G. D. (1970). Representation and retrieval processes in short-term memory. *Journal of Experimental Psychology, 10*, 438–464.

Smith, E. E., Rips, L. J., & Shoben, E. J. (1974). Semantic memory and psychological semantics. In G. H. Bower (Ed.), *The psychology of learning and motivation* (Vol. 8, pp. 1–45). New York: Academic Press.

Smoking and health report of the advisory committee to the Surgeon General of the Public Health Service. (1964). Washington, DC: U.S. Department of Health, Education, and Welfare.

Sowder, J. (1992). Estimation and number sense. In D. A. Grouws (Ed.), *Handbook on research in mathematics teaching and learning* (pp. 371–389). New York: Macmillan.

Staley, D. J. (2003). *Computers, visualization, and history: How new technology will transform our understanding of the past*. Armonk, NY: M. E. Sharpe.

Stanfield, R. A., & Zwaan, R. A. (2001). The effect of implied orientation derived from verbal context on picture recognition. *Psychological Science, 12*, 153–156.

Sternberg, R. J., & Ben-Zeev, T. (2001). *Complex cognition: The psychology of human thought*. New York: Oxford University Press.

Stevenson, R. L. (1895/1993). The strange case of Dr. Jekyll and Mr. Hyde. In I. Bell (Ed.), *Robert Louis Stevenson: The complete short stories* (Vol. 2, pp. 102–164). Edinburgh: Mainstream Publishing.

Sumin, G. R., & Smith, E. R. (Eds.). (2008). *Embodied grounding: Social, cognitive, affective, and neuroscientific approaches*. Cambridge, England: Cambridge University Press.

Svenson, O. (2009). Driving speed changes and subjective estimates of time savings, accident risks and braking. *Applied Cognitive Psychology, 23*, 543–560.

Swartout, W., Gratch, R. H., Hill, R., Hovy, E., Marsella, S., Rickel, J., et al. (2006). Toward virtual humans. *AI Magazine, 27*, 96–108.

Sweller, J. (2003). Evolution of human cognitive architecture. In B. Ross (Ed.), *The psychology of learning and motivation* (Vol. 43, pp. 215–266). San Diego, CA: Academic Press.

Sweller, J., & Chandler, P. (1994). Why some material is difficult to learn. *Cognition and Instruction, 12*, 185–233.

Tater, D., Roschelle, J., Knudsen, J., Shechtman, N., Kaput, J., & Hopkins, B. (2008). Scaling up innovative technology-based mathematics. *The Journal of the Learning Sciences, 17*, 248–286.

Thompson, P. W. (1994). Concrete materials and teaching for mathematical understanding. *Arithmetic Teacher, 40*, 556–558.

Trickett, S. B., & Trafton, J. G. (2007). "What if." The use of conceptual simulations in scientific reasoning. *Cognitive Science, 31*, 843–875.

Triona, L. M., & Klahr, D. (2003). Point and click or grab and heft: Comparing the influence of physical and virtual instructional materials on elementary school students' ability to design experiments. *Cognition and Instruction, 21*, 149–173.

Tufte, E. R. (2001). *The visual display of quantitative information* (2nd ed.). Cheshire, CT: Graphics Press.

Turati, C., Bulf, H., & Simion, F. (2008). Newborns' face recognition over changes in viewpoint. *Cognition, 106*, 1300–1321.

Tversky, B. (2005). Visuospatial reasoning. In K. J. Holyoak & R. G. Morrison (Eds.), *The Cambridge handbook of thinking and reasoning* (pp. 209–239). Cambridge, England: Cambridge University Press.

Tversky, B., Morrison, J. B., & Betrancourt, M. (2002). Animation: Can it facilitate? *International Journal of Human-Computer Studies, 57*, 1–16.

U.S. Department of Commerce. (2002). Visions 2020: Transforming Education and Training through Advanced Technologies. Washington, DC: Department of Commerce.

U.S. Department of Commerce. (2005). Visions 2020.2 Student Views on Transforming Education and Training through Advanced Technologies: Department of Commerce.

U.S. Department of Education National Commission on Excellence in Education. (1983). A National Risk. http://www.ed.gov/pubs/NatAtRisk/risk.html.

Van Meter, P. (2001). Drawing construction as a strategy for learning from text. *Journal of Educational Psychology, 69*, 129–140.

Van Meter, P., & Garner, J. (2005). The promise and practice of learner-generated drawing: Literature review and synthesis. *Educational Psychology Review, 17*, 285–325.

Vekiri, I. (2002). What is the value of graphical displays in learning? *Educational Psychology Review, 14*, 261–312.

Verschaffel, L., Greer, B., & De Corte, E. (2000). *Making sense of world problems*. Lisse, The Netherlands: Swets & Zeitling B.V.

Vygotsky, L. (1962). *Thought and language*. Cambridge, MA: The MIT Press.

Wagenaar, W. A., & Sagaria, S. D. (1975). Misperception of exponential growth. *Perception & Psychophysics, 18*, 416–422.

References

Wagner, S. M., Nusbaum, H., & Goldin-Meadow, S. (2004). Probing the mental representation of gesture: Is handwaving spatial? *Journal of Memory and Language, 50*, 395–407.

Walsh, V. (2003). A theory of magnitude: Common cortical metrics of time, space, and quantity. *TRENDS in Cognitive Sciences, 7*, 483–488.

Watson, J. B. (1924). *Behaviorism*. New York: Norton.

Webb, R. M., Lubinski, D., & Benbow, C. P. (2007). Spatial ability: A neglected dimension in talent searches for intellectually precocious youth. *Journal of Educational Psychology, 99*, 397–420.

Weise, H. (2003). Iconic and non-iconic stages in number development. *TRENDS in Cognitive Sciences, 7*, 385–390.

Wertheimer, M. (1959). *Productive thinking*. New York: Harper & Row.

Wheeler, M. E., & Reed, S. K. (1975). Response to before and after Watergate caricatures. *Journalism Quarterly, 52*, 134–137.

White, B. Y. (1984). Designing computer games to help physics students understand Newton's laws of motion. *Cognition and Instruction, 1*, 69–108.

White, B. Y. (1993). ThinkerTools: Causal models, conceptual change, and science education. *Cognition and Instruction, 10*, 1–100.

Wieman, C. E., Adams, W. K., & Perkins, K. K. (2008). PhET: Simulations that enhance learning. *Science, 322*, 682–683.

Wilensky, U., & Reisman, K. (2006). Thinking like a wolf, a sheep, or a firefly: Learning biology through constructing and testing computational theories—an embodied modeling approach. *Cognition and Instruction, 24*, 171–209.

Wilson, M. (2002). Six views of embodied cognition. *Psychonomic Bulletin & Review, 9*, 625–636.

Wu, H.-K., Krajcik, J., & Soloway, E. (2001). Promoting understanding of chemical representations: Students' use of a visualization tool in the classroom. *Journal of Research in Science Teaching, 38*, 821–842.

Wu, H.-K., & Shah, P. (2004). Exploring visuospatial thinking in chemistry learning. *Science Education, 88*, 465–492.

Yerushalmy, M. (1997). Mathematizing verbal descriptions of situations: A language to support modeling. *Cognition and Instruction, 15*, 207–264.

Young, M. (2004). An ecological psychology of instructional design: Learning and thinking by perceiving-acting systems. In D. H. Jonassen (Ed.), *Handbook of research for educational communications and technology* (2nd ed., pp. 169–177). Mahwah, NJ: Lawrence Erlbaum Associates.

Zachry, M., & Thralls, C. (2004). An interview with Edward R. Tufte. *Technical Communication Quarterly, 13*, 447–462.

Zwaan, R. A., & Yaxley, R. H. (2003). Spatial iconicity affects semantic relatedness judgments. *Psychonomic Bulletin & Review, 10*, 954–958.

Author Index

A

Aberg, C., 126, 189
Adams, V. M., 192
Adams, W. K., 191
Ahl, V. A., 30, 34, 184
Akaygun, S., 158, 191
Allen, B. S., 138, 140, 190
Anglin, G. J., 79, 83, 187
Ardac, D., 158, 191

B

Baddeley, A. D., 113, 114, 189
Bainbridge, W. S., 190
Baker, E., 174
Ball, T. M., 59, 61, 62, 186
Barbey, A. K., 7, 183
Barsalou, L. W., 7, 6, 10, 17, 47, 173, 183
Barton, B., 192
Barwell, R., 192
Benbow, C. P., 192
Bentall, R. P., 56, 186
Ben-Zeev, T., 1, 183
Bergen, B. K., 189
Bertheau, M., 29, 34, 42, 184
Betrancourt, M., 187
Biswas, G., 188
Black, J. B., 66, 83, 186
Black, T., 129, 130, 190
Blair, K. P., 188
Bongard, J., 190
Booth, J. L., 27, 28, 184
Boroditsky, L., 22, 184
Bowers, J. S., 192
Brandimonte, M. A., 54, 186
Brezinski, K. L., 27, 28, 34, 184
Bright, G. W., 99–101, 188
Brooks, L. R., 49, 186
Brown, D., 4, 183
Brunye, T. T., 10, 189
Bulf, H., 183
Burkhardt, H., 193

C

Calvo, M. G., 75, 187
Campbell, A. E., 192
Campbell, R. L., 132, 190
Capote, T., 3, 35, 183
Carey, S., 17, 184
Carlson, M., 102, 103, 104, 105, 188
Carroll, J. M., 91, 198
Casasanto, D., 22, 184
Cate, C., 83, 84, 187
Cemen, P. B., 41, 185
Chambers, D., 54, 186
Chandler, P., 115, 116, 189
Chase, W. G., 128, 190
Chatterjee, A., 17, 183–184
Chen, W., 8, 183
Chiu, J. L., 160, 191
Clark, R. C., 174, 192
Clements, D. H., 132, 190
Cobb, S., 135, 190
Coe, E., 102, 103, 104, 105, 188
Cohen, J., 111
Cohen, L., 184
Cole, K. C., 106, 108, 188
Collins, A. M., 96, 188
Conroy, P., 2, 183
Cooke, J., 192
Cornell, E., 14, 15, 76, 183
Cremer, J. F., 143, 191
Cunningham, K. L., 79, 83, 187
Curcio, F. R., 99–101, 188

D

D'Agostino, P. R., 122, 123, 189
Dansereau, D. F., 96–97, 188
Dansereau, D. G., 97, 188
Davis, G. E., 192
Davis, J., 188
Davis, R., 10, 183
Dawson, M., 186
De Corte, E., 165
de Groot, A. D., 128, 190
Dehaene, S., 18, 20, 22, 23, 28, 184
DeLoache, J., 4, 183
Dickieson, J., 174
Diezmann, C., 89, 90, 187
Difede, J., 142
diSessa, A. A., 88, 187
Dixon, J. A., 30, 31, 32, 34, 184
Doll, R., 102, 188
Dove, G., 188

E

Edwards, L. D., 185
Engelkamp, J., 125, 189
English, L. D., 89, 90, 187
Engstler-Schooler, T. Y., 54, 186

Etschenberg, S., 56, 186
Evans, A. C., 34, 185

F

Fagan, J. F., 13, 14, 76, 183
Fantz, R., 13, 14, 76, 183
Feigenson, L., 17, 184
Fincher-Kiefer, R., 122, 123, 189
Finke, R. A., 68, 70, 71, 72, 186
Fisher, D. L., 144, 191
Fisher, K. M., 98, 188
Forbur, K. D., 79, 81
Francis, M., 93, 187–188
Fraser, D. S., 135, 190
Friedman, M. P., 51, 186
Friel, S. N., 99–101, 188

G

Gabriele, A. J., 29, 34, 42, 184
Gahm, G., 142
Gallistel, C. R., 184
Ganis, G., 185
Garber, P., 190
Garner, J., 187
Gelman, R., 29, 34, 42, 184
Gerbino, W., 54, 186
Gernsbacher, M. A., 186
Gibbons, K., 3, 35, 183
Gibbs, R. W., 183
Gibson, J. J., 137, 190
Gleitman, L., 183
Glenberg, A. M., 120–121, 133, 189, 190
Goldin-Meadow, S., 130, 131, 190
Goldstone, R. L., 153, 154, 155, 191
Gorodetsky, M., 98, 188
Grandin, T., 65, 66, 186
Grant, E. R., 82, 83, 187
Gratch, R. H., 142, 191
Greer, B., 165, 166, 192
Grubb, S., 190
Gruneberg, M., 19
Gutierrez, T., 120–121, 133, 189

H

Haider, H., 68, 186
Hajela, D., 187
Hall, R. F., 97, 188
Hampson, P., 48, 185
Harp, S. F., 78, 79, 187
Hauser, M., 17, 184
Hauser, M. D., 18, 184
Hayadrw, E. O., 187
Hebert, J., 190

Hegarty, M., 64, 83, 84, 89, 90, 102, 103, 186, 187, 188
Heidbreder, E., 185
Heiser, J., 183
Herrman, D. J., 19
Hesslow, G., 183
Hidi, S., 78, 187
Hill, R., 142, 191
Hinrichs, T. R., 79, 81
Hirshberg, J., 72, 186
Hitch, G., 113, 189
Hock, H., 61, 62, 63, 186
Hoffler, T. N., 187
Hoffman, B., 84, 85, 109, 111, 138, 140, 166, 168, 187, 188, 190, 192
Hogan, T. P., 27, 28, 34, 184
Holley, C. D., 96–97, 188
Holyoak, K. J., 190
Homer, B. D., 187
Hopkins, B., 171, 192
Horton, W. S., 119, 120, 189
Hovy, E., 142, 191
Hsu, E., 102, 103, 104, 105, 188
Hurley, S. M., 93, 94, 187–188, 188
Husic, F., 160, 191
Hyman, I. E., 57, 186

J

Jacobs, S., 102, 103, 104, 105, 188
Jahn, G., 123, 189
Japuntich, S., 120–121, 133, 189
Jarvin, L., 190
Jaworski, B., 121, 189, 190
Jazo, L., 192
Johnson, C., 65, 66, 186
Johnson, L. F., 176, 193
Johnson, M., 35, 36, 46, 183, 185
Johnson, M. K., 55, 56, 186
Jones, S., 48, 185
Jones, T., 157, 158, 191
Joram, E., 29, 34, 42, 184

K

Kaput, J., 171, 192
Kaput, J. J., 161, 162, 163, 171, 192
Kaschak, M. P., 120–121, 133, 189
Kato, T., 8, 183
Katz, V. J., 192
Kearney, J. K., 143, 191
Kiesel, A., 184
Kintsch, W., 170, 192
Klahr, D., 159, 160, 191
Klein, G. A., 141, 145, 150, 174, 190
Klingberg, T., 126, 189

Author Index

Knoblich, G., 68, 186
Knudsen, J., 171, 192
Kohler, W., 67, 186
Koriat, A., 125, 126, 189
Kosslyn, S. M., 59, 61, 62, 185, 186
Kozhevnikov, M., 89, 90, 102, 103, 187, 188
Kozma, R., 157, 158, 191
Krajcik, J., 156, 157, 191
Kriz, S., 83, 84, 187

L

Lafleur, L. J., 45
Lakoff, G., 35, 36, 37, 38, 39, 40, 41, 44, 46, 47, 183, 185
Landauer, T. K., 21, 184
Lang, P. J., 75, 187
Larsen, S., 102, 103, 104, 105, 188
Lee, H., 160, 191
Leeawong, K., 188
Leutner, D., 187
Levin, J. R., 78, 120–121, 133, 187, 189, 190
Lieberman, B., 193
Linden, D. E., 56, 186
Lindsay, D. S., 56, 57, 186
Lindsay, S., 189
Linn, M. C., 160, 191
Lobato, J., 79, 80, 187
Lockhead, G. R., 61, 62, 63, 186
Loftus, E. F., 56, 96, 186, 188
Lorayne, H., 48, 185
Lubinski, D., 192
Lucas, J., 48, 185

M

Malhotra, A., 91, 198
Mandler, J., 15, 16, 17, 183
Mandler, J. M., 16, 183
Markman, A. B., 183
Marsella, S., , 191
Martin, T., 130, 190
Martin, W. O., 107, 188
Marx, N., 157, 158, 191
Matlock, T., 41, 185, 189
Matos, J. F., 185
Mayer, R. E., 78, 79, 102, 117, 118, 119, 123, 187, 188, 189
McDonough, L., 16, 183
McNeil, N. M., 190
Metcalf, J., 67, 186
Metzler, J., 59, 60, 186
Meyer, R., 187
Miller, A. I., 183
Mitchell, D. B., 60, 186
Molko, N., 184

Moore, C. F., 30, 31, 32, 34, 184
Moran, C., 192
Moreno, R., 187
Morris, P. E., 48, 185
Morrison, J. B., 187
Mottron, L., 186
Moyer, P. S., 133, 190
Moyer, R. S., 21, 184

N

Narayanan, S., 189
Nasir, N. J., 130, 190
Nathan, M. J., 170, 192
National Science Board Commission on Precollege Education in Mathematics, Science, and Technology, 175, 192
Neisser, U., 36, 136, 137, 138, 139, 144–145, 150, 174, 185, 190
Nemirovsky, R., 192
Newell, A., 42, 185
Nickerson, S. N., 192
Nielsen, G. D., 50, 186
Nilsson, L.-G., 126, 189
Noice, H., 126, 127, 128, 189
Noice, T., 126, 127, 128, 189
Norretranders, T., 187
Novick, L. R., 93, 94, 187–188, 188
Nuñez, R. E., 37, 38, 39, 40, 41, 44, 47, 177, 185, 193
Nusbaum, H., 130, 190
Nyberg, L., 126, 189
Nydam, C., 192

O

O'Donnell, A. M., 97, 188
Ogawa, S., 8, 183
Ohlsson, S., 68, 186
O'Neil, H. F., 174
Opfer, J. E., 26, 34, 42, 184
Otto, R. G., 138, 140, 190

P

Paivio, A., 47, 48, 49, 125, 185
Palmer, S. E., 183
Pani, J. R., 47, 53, 185
Papafragou, A., 183
Papert, S., 147, 191
Payne, D. G., 19
Pearlman-Avnion, S., 125, 126, 189
Pentland, J., 57, 186
Perkins, K. K., 191
Persson, J., 126, 189

Pesce, M., 135, 190
Pfeifer, R., 190
Phares, S., 168, 192
Philipp, R. A., 107, 188
Piaget, J., 16, 131, 139, 174, 190
Pinel, P., 22, 23, 184
Pinker, S., 3, 5, 183
Plass, J. L., 187
Plumert, J. M., 143, 191
Pollatsek, A., 144, 191
Pomerantz, J. R., 185
Pradhan, A., 144, 191
Pugh, H. L., 139, 141, 191
Pylyshyn, Z., 59, 185, 186

R

Ramani, G., 27, 34, 184
Randel, J. M., 139, 141, 191
Rapp, D. N., 119, 120, 189
Raye, C. L., 55, 56, 186
Read, D. D., 56, 57, 186
Reed, S. K., 34, 36, 37, 43, 44, 51, 52, 61, 62, 63, 76, 77, 84, 85, 109, 111, 139, 141, 166, 168, 185, 186, 187, 188, 191, 192
Reger, G., 142
Reisberg, D., 54, 186
Reiser, B. J., 59, 61, 62, 186
Reisman, K., 151, 152, 191
Renninger, K. A., 78, 187
Resnick, M., 151, 191
Rhenius, D., 68, 186
Richgels, G. W., 107, 188
Richland, L. E., 190
Richman, C. L., 60, 186
Rickel, J., 142, 191
Rieber, L. P., 83, 147, 187, 191
Rips, L. J., 45, 185
Rischal, M., 121, 189, 190
Rizzo, A., 142
Roland, P. E., 126, 189
Roschelle, J., 161, 162, 163, 171, 192
Roth, M., 190
Rothbaum, B. O., 142
Rubin, D. C., 183
Russell, J., 157, 158, 191

S

Saariluoma, P., 114, 189
Sack, A. T., 56, 186
Sagaria, S. D., 188
Sarama, J., 190
Schoenfeld, A. H., 193
Schooler, J. W., 54, 186

Schwartz, D. L., 66, 83, 129, 130, 183, 186, 188, 190
Schwarz, C. V., 150, 191
Sedig, K., 192
Setati, M., 192
Shah, P., 156, 157, 191
Shechtman, N., 171, 192
Shepard, R. N., 59, 60, 183, 185, 186
Shoben, E. J., 45, 185
Short, D., 109, 111, 188
Shrobe, H., 183
Shu, X. H., 8, 183
Siegler, R. S., 26, 27, 28, 34, 42, 184
Simion, F., 183
Simmons, W. K., 7, 183
Simon, H. A., 42, 43, 44, 128, 185, 190
Singer, J. L., 56, 186
Smith, E. E., 45, 50, 185, 186
Smith, E. R., 183
Smith, R. S., 176, 193
Smith, S. M., 68, 72, 186
Soloway, E., 156, 157, 191
Son, J. Y., 153, 154, 155, 191
Soulieres, I., 186
Sowder, J., 184
Spelke, E., 22, 23, 184
Spivey, M. J., 82, 83, 187
Staines, G., 128, 189
Staley, D. J., 190
Stanescu, R., 22, 23, 184
Stanfield, R. A., 8, 9, 119, 183
Sternberg, R. J., 1, 183
Stevenson, R. L., 120, 189
Stroup, W., 161, 162, 163, 171, 192
Subrahmanyam, K., 29, 34, 42, 184
Sumin, G. R., 183
Sumner, M., 192
Svenson, O., 144
Swartout, W., 142, 191
Sweetser, E., 177, 193
Sweller, J., 115, 116, 117, 123, 158, 189
Szolovits, P., 183

T

Tank, D. W., 8, 183
Tater, D., 171, 192
Taylor, H. A., 10, 189
Taylor, T. T., 55, 186
Thomas, J. C., 91, 198
Thompson, P. W., 132, 190
Thompson, W. L., 185
Thralls, C., 188
Tierney, C., 192
Tinker, R., 160, 191
Trafton, J. G., 188

Author Index

Trickett, S. B., 188
Triona, L. M., 159, 160, 191
Tsivkin, S., 22, 23, 184
Tufte, E. R., 99, 100, 166, 188
Turati, C., 183
Tversky, B., 183, 187

U

Ugurbil, K., 8, 183
U.S. Department of Commerce, 176, 193
U.S. Department of Education National Commission on Excellence in Education, 175
U.S. Department of Health, Education, and Welfare, 188

V

Vaez, H., 79, 83, 187
van de Ven, V. G., 56, 186
Van Meter, P., 80, 187
Vekiri, I., 187
Verschaffel, L., 165
Vierck, E., 184
Vygotsky, L., 183

W

Wagenaar, W. A., 188
Wagner, S. M., 130, 190
Walsh, V., 18, 19, 20, 184
Wang, A. Y., 5, 186
Ward, T. B., 68, 72, 186
Watson, J. B., 185
Webb, R. M., 192
Weise, H., 184
Wertheimer, M., 186
Wheeler, M. E., 76, 77, 187
White, B. Y., 148, 149, 150, 155, 191
Wieman, C. E., 191
Wilensky, U., 151, 152, 191
Williams, C., 191
Wilson, A. J., 184
Wilson, C. D., 7, 183
Wilson, M., 6, 183
Wright, T., 192
Wu, H.-K., 156, 157, 191
Wulfeck, W., 174

Y

Yaxley, R. H., 9, 183
Yerushalmy, M., 104, 105, 188
Yoder, C. Y., 19
Young, E., 170, 192
Young, M., 137, 150, 190

Z

Zachry, M., 188
Zur, O., 190
Zwaan, R. A., 8, 9, 119, 183

Subject Index

A

Acid Problem, 33
Action, vision and, 125–134, 189
    abstract manipulation group, 133
    acting, 126–128
    active experiencing, 126, 127
    chess, 128–129
    chunks, 128
    empirical abstraction, 132
    gesturing while explaining solution, 131
    information mismatch, 131
    knowledge revealed through action, 130
    manipulatives, 132–134
    memory for actions, 125–126
    motor memory codes, 126
    reasoning, 129–131
    reflecting abstraction, 132
    reflections, 131–132
    semantic codes, creation of, 126
    story-relevant manipulation group, 133
    summary, 134
    working memory, cognitive demands on, 130
Active experiencing, 126, 127
Algebra Sketchbook, 104–106
Alice's Adventures in Wonderland, 87
Ambiguous figures, 54, 186
American functionalism, 47
Amodal knowledge, 173
Amodal symbols, 95
    systems, 6, 7
    visualization and, 96
Animals in Translation, 65, 186
ANIMATE, 169–171, 174, 192
Animated pictures, 82–84
Animation Tutor™ DVD, 164, 179–182
    average speed, 181
    catch up, 181
    chemical kinetics, 180
    dimensional thinking, 179–180
    leaky tanks, 182
    personal finance, 180
    population growth, 180–181
    task completion, 181–182
Architecture, 1–3
Autism, 65, 186
Average Speed Problem, 167

B

Bacteria Problem, 106

Behaviorism, 47
Bodily motion, conversation and, 130
Bottom–up processing, 37
Brain imaging, 6, 8, 23
Braking Problem, 168
Bridge Problem, 167
Bridging inference, 122

C

Causal reasoning, 64–66, 80
Chemistry, instructional software, 156–159
Chess, vision and action, 128–129
Clown Problem, 161–162
Cognition
    embodied, 133, 173, 177, 183
    goal-directed, 47
    perception working against, 25
    verbal, 1
    visual, 11
Cognition and Reality, 136, 137, 139
Cognitive abilities, 13, *see also* Images before words
Cognitive development, prevalent intuition, 16
Cognitive effect, picture viewing, 76
Cognitive load theory, 115, 189
Cognitive operations, memory for, 55
Cognitive overload, 115, 116
Cognitive processes
    central, 6
    visual imagery and, 47
Cognitive Psychology, 136, 185
Cognitive psychology, 35, 36
Cognitive revolution, 53
Cognitive simulation, 41
Cognitive skills
    finding relations in data, 101
    graph comprehension, 99
    reading beyond data, 102
    reading of data, 101
Coherence principle, 118
Companion Cognitive Systems, 79, 81
Complex Cognition: The Psychology of Human Thought, 1
Computational task, 28
Computer models, visualization and, 150
Computers, Visualization, and History, 190
Conceptual thought, origins of, 16
Concrete information, 151
Containment image-schema, 16
Contextual information, 55

215

216 Subject Index

Creative Cognition: Theory, Research, and Applications, 186
Creative Imagery: Discoveries and Inventions in Visualization, 186
Creative Priority, The, 72, 186

D

Da Vinci Code, The, 4, 183
Decorative pictures, 76–78
Diagram production, 87–98, 187
   amodal symbols, visualization and, 96
   constructing representations, 88–89
   diagram of frog problem, 89
   diagram of koala problem, 90
   diagram of motion, student's, 88
   emphasizing spatial relations, 89–90
   example problems, 93
   hierarchical relations, 92
   hierarchy, 93
   knowledge organization, 96, 97
   matching diagrams to problems, 92–95
   matrix, 91, 92, 93
   network, 93
   Paraval Problem, 93, 94, 95
   Path Problem, 90
   Project MaRC, 88
   puzzle solving, 87
   representations, judging of, 88
   representing space and time, 91–92
   semantic network model of conceptual organization, 96
   semantic networks as instructional tools, 96–97
   SemNet network, 98
   spatial representations of meaning, 95–96
   spatial version, 91
   summary, 98
   Venn diagram, 93, 95
   visual thinking, basis for, 95
Dimensional Thinking module, 165
Discourse on Method, 45
DNA models, 1

E

eChem, 156, 191
Ecological psychology, virtual reality and, 137
Ecological systems, instructional software, 150–153
Embodied cognition, 133, 173, 177, 183
Empirical abstraction, 132
Estimation, 25–34, 184
   Acid Problem, 33
   approximations, 28
   computational task, 28
   Fechner's law, 25
   interpolation, 33–34
   measurement, 27, 28–30
   mental number line, 26–27
   mixture problems, 30–33
   numerosity task, 27
   principal components analysis, 28
   quantity principle, 30
   range principle, 30
   spatial skills, 27–28
   summary, 34
   tasks, dot displays used in, 20
   Temperature Problem, 30
Evocative pictures, 75–76
Expertise reversal effect, 117
Exploration strategy, 72

F

Fallingwater, 1, 3
Fechner's law, 25
Fish problem, 163
Foundations of Mind: Origins of Conceptual Thought, The, 15, 183

G

Galloping Horse Problem, 122
Gardener's Problem, 67
Geneplore model, 71–72
Generation strategy, 71
Gestalt psychology, 66, 67
Gestures, reasoning and, 130
Graph comprehension, 99–111, 173, 188
   Algebra Sketchbook, 104–106
   Animation Tutor, 109, 111
   Bacteria Problem, 106
   exponential growth, 106–108
   graphs of linear and exponential growth, 107
   interpreting events over time, 102–103
   modeling population growth, 108–111
   Napoleon's march to Moscow, 100
   outliers, 102
   Pipeline Problem, 104
   Pollution Problem, 107, 108
   reasoning from graphs, 99–102
   relation between cigarette consumption and death from lung cancer, 101
   Salary Problem, 107
   student interpretation, 102
   summary, 111
   Temperature Problem, 103, 104
Grounding metaphors, 40–42

Subject Index

H

Hallucinations, 56–57
Handbook of Research for Educational Communications and Technology, 79, 137
Hollyhock House, 2
Horizon Report, The, 176

I

Image manipulation, 59–73, 186
  arithmetic operations, 68
  causal reasoning, 64–66
  designing with images, 68–71
  equal signs, 68
  exploration strategy, 72
  Geneplore model, 71–72
  generation strategy, 71
  insight, definition of, 67
  map of island, 61
  mental scanning, 62
  model of transition from simulations to rules, 66
  moment of illumination, 67
  multiple interpretations of single preinventive form, 71
  object parts used to construct inventions, 70
  objects as preinventive forms, 69
  pairs of complex patterns differing in orientation, 60
  problem-solving tasks, 73
  pulley problem, 64
  rearranging objects, 66–68
  scanning images, 59–64
  scan times for perceived patterns, 63
  summary, 73
Image production, 47–58, 185
  advantages of images, 49–51
  ambiguous figure, 54
  American functionalism, 47
  breakdown of reality monitoring, 56–57
  categories of objects, 51
  cognitive operations, memory for, 55
  contextual information, 55
  dual coding theory, 48, 185
  elaboration, forms of, 48
  high-imagery words, 48
  limitations of images, 53–55
  memory codes, 50
  object categories, 51
  perceptual symbols system, 47
  prototype creation, 52
  prototypical images, 51–53
  reality monitoring, 55–56
  remembering visual images, 47–49
  schematic faces, 51
  sensory information, 55
  summary, 57–58
  test faces for categorization, 52
  verbal overshadowing effect, 54
Images before words, 13–23, 183
  cerebral cortex, 19
  cognitive abilities, 13
  cognitive development, prevalent intuition, 16
  containment image-schema, 16
  estimation tasks, dot displays used in, 20
  experimental study of infants, 13
  memory to concepts, 14–17
  mental simulation, 17
  Number Comparison Task, 21
  numerical reasoning with language, 22–23
  origins of conceptual thought, 16
  path image-schema, 16
  perception, 17
  small quantities, 17–18
  space, quantity, and time, 18–22
  summary, 23
  theory of magnitude, 18, 19
  verbal representation, 17
  visual memory, 13–14
Images versus words, 1–12, 183
  achievements of humanity, 1
  amodal symbol systems, 6, 7
  book organization, 10–11
  brain imaging, 6, 8
  DNA models, 1
  examples of visual thinking, 1
  Frank Lloyd Wright designs, 2
  language, richness of, 5
  language of thought, 4
  mental simulations, 1
  Object Verification Task, 9
  perceptual symbols, 6–10
  propositions, 5
  psycholinguistics, 1
  spatial visualization, 1
  summary, 11–12
  symbol intention, 4
  thinking with symbols, 3–5
  visualization, action and, 11
  visualization in scientific discovery, 1
  visual metaphor, making abstract concrete, 2
  visual simulation hypothesis, 9
  visual thinking, words reflecting power of, 1
  Word Association Task, 9
Image viewing, *see* Viewing of pictures

Subject Index

In Cold Blood, 3
Individual differences principle, 118
Information
　abstract, mind as processor of, 6
　abstraction of, 132
　cognitive overload, 115
　concrete, 151
　contextual, 55
　flow, stages, 185
　inability to transfer, 79
　mismatch, 131
　optical, 137
　perceptual, 97
　phonological, 114
　-processing models, 136, 138
　redundancy effect, 116, 119
　representation, 7
　sensory, 36, 55
　unreliability, 139
　verbal, 113
　visual, high-resolution, 143
　visual image vs. verbal code, 50
Insight, definition of, 67
Inventions, object parts used to construct, 70

J

Johnson Wax building, 2

K

Knowledge
　abstract form of, 6
　amodal, 173
　democratization of, 161
　generalized, 27
　microworlds and, 148
　military, 142
　modeling of, 5
　operative, 132
　organization, 96, 97
　perceptual theory of, 6
　prior, LTM and, 113, 118
　problem diagrams, 93
　procedural-motor, 187
　representation, 1, 183
　revealed through action, 130
　transmission of, 4

L

La Geometrie, 45
Language
　from perception to, 17
　numerical reasoning with, 22-23
　richness of, 5
　spatial ability and, 13
　of thought, 4
Language Instinct, The, 3
Life All Around Me by Ellen Foster, The, 3
Logo programming language, 147
Long-term memory (LTM), 37, 114
LTM, *see* Long-term memory

M

Magnitude system, 18, 19
Making Sense of Word Problems, 165
Manipulation of images, *see* Image
　　manipulation
Maps, 10, 61, 87, 100
MaRC group, 88
Matchstick problems, 68
Mathematical reasoning, 38, 84
Mathematics instructional software, 161-171,
　　192
　ANIMATE, 169-171
　Animation Tutor, 164, 166, 168
　Average Speed Problem, 167
　Braking Problem, 168
　Bridge Problem, 167
　Clown Problem, 161-162
　democratization of knowledge, 161
　fish problem, 163
　object manipulation, 165-167
　Pizza Problem, 165
　SimCalc, 161-164
　simulation, 167-169
　summary, 171
　visualization, school curricula supporting,
　　171
Means-end strategy, 44
Measurement task, 27
Memory
　codes, image production and, 50
　verbal, images and, 49
　vision and actions, 125
　visual, 13-14
　working, model of, 114
Memory for Actions, 125, 189
Mentality of Apes, The, 67
Mental scanning, 62
Metaphors, *see* Spatial metaphors
Metaphors We Live By, 35, 46, 185
Microworlds, 147, 160
Military knowledge, 142
*Mindstorms: Children, Computers, and
　　Powerful Ideas*, 191
Missing Dollar Problem, 38-39
Missionaries-Cannibals Problem, 42, 43

Subject Index

Modality principle, 118
Model(s)
    computer, visualization and, 150
    DNA, 1
    Geneplore, 71–72
    information-processing, 136, 138
    multimedia, 117
    recognition-primed, 141, 150
    topic interest, 78
    transitions from simulations to rules, 66
    working memory, 114
Model-Enhanced ThinkerTools curriculum, 150
Moment of illumination, 67
Motion, student's diagram of, 88
Motor memory codes, 126
Multidimensional scaling, 45
Multimedia
    instruction, design principles, 118
    learning, 174–175
    model, 117
    principle, 118
MultiMedia and Mental Models software, 157, 158

N

National Science Board, recommendations, 175
Negotiation skills, training of, 142
NetLogo, 151
Number Comparison Task, 21
Number Sense: How the Mind Creates Mathematics, The, 18, 184
Numerosity task, 27

O

Object(s)
    manipulation, mathematics instructional software, 165
    memory for, 14
    parts used to construct inventions, 70
    as preinventive forms, 69
Object Verification Task, 9
Operative knowledge, 132

P

Path image-schema, 16
Path Problem, 90
Pattern recognition, 36
Perception, *see also* Images before words
    condition, 61, 62
    construction view of, 136=137
    from perception to language, 17
    Gestalt interest in, 67
    imagination and, 55
    infants, 13
    theory of, 137
Perceptual cycle, 138
Perceptual symbols, 6–10
    hypothesis, 173
    system, 47
Physics, instructional software, 148–150
Pictures, *see* Viewing of pictures; Words and pictures
Pipeline Problem, 104
Pizza Problem, 165
Playful World: How Technology is Transforming Our Imagination, The, 135, 190
Political cartoons, 76
Pollution Problem, 107, 108
Pong, 135
Predictions, 177
Prince of Tides, The, 2
Principal components analysis, 28
Problem(s)
    Acid, 33
    Average Speed, 167
    Bacteria, 106
    Braking, 168
    Bridge, 167
    Clown, 161–162
    fish, 163
    Galloping Horse, 122
    Gardener's, 67
    matching diagrams to, 92
    matchstick, 68
    mathematical, visualization of, 90
    matrix, 91
    Missing Dollar, 38–39
    Missionaries–Cannibals, 42, 43
    Paraval, 93, 94, 95
    Path, 90
    Pipeline, 104
    Pizza, 165
    Pollution, 107, 108
    pulley, 64, 83
    Radiation, 82
    Salary, 107
    Temperature, 30
Procedural-motor knowledge, 187
Programming
    course, characters used in, 136, 190
    language, Logo, 147
Project MaRC, 88
Proportional reasoning, 30, 85, 84, 168, 180
Propositional representations, 5
Prototype creation, 52
Psycholinguistics, 1

Puzzle solving, 87

Q

Quantity principle, 30

R

Radiation Problem, 82
Range principle, 30
Reality monitoring, 55
Reasoning
 causal, 64, 80
 gestures and, 130
 graphs, 99–102
 mathematical, 38, 84
 numerical, 18, 22
 proportional, 30, 84, 85, 168, 180
 spatial, 34
 tasks, 129
 vision and action, 129–131
Recherches sur l'abstraction réfléchissante, 139
Recognition-primed model, 141, 150
Recommendations, 175–176
 National Science Board, 175
 U.S. Department of Commerce, 175
 U.S. Department of Education, 175
Redundancy effect, 116, 118, 119
Reflecting abstraction, 132
Robie House, 2
Roman numerals, 68
Rules, transitions from simulations to, 66

S

Salary Problem, 107
Schizophrenia, 56
Science instructional software, 147–160, 191
 advantages of concrete and idealized representations, 155
 ants and food simulation, 154
 chemistry, 156–159
 common misconception, 148
 computer models, 150
 design of experiments, 160
 eChem, 156, 191
 ecological systems, 150–153
 green dot, 153
 inquiry-based science curriculum, 156
 losers, 153
 microworlds, 147, 160
 MultiMedia and Mental Models, 157, 158
 NetLogo, 151
 outcomes from varying parameter values, 152
 physics, 148–150
 scientific design, 159–160
 StarLogo, 151
 summary, 160
 ThinkerTools software, 149
 transfer of principles, 153–155
 winners, 153
Scientific discovery, 1–2
Search spaces, 42–44
Semantic codes, creation of, 126
Semantic networks, 95
SemNet network, 98
Sensory information, 55
Sensory store, 36
Seven Psychologies, 185
Shapes program, 132
Short-term memory (STM), 36, 113
SimCalc, 161–164
SimCity, 135
Situation awareness, virtual reality and, 139–141
Software, see Mathematics instructional software; Science instructional software
Source–path–goal schema, 185
Spatial contiguity principle, 118
Spatial metaphors, 35–46, 173, 185
 applications to cognitive psychology, 36–37
 applications to mathematics, 37–40
 balance strategy, 44
 bottom–up processing, 37
 filter, 36
 grounding metaphors, 40–42
 linking metaphors, 45–46
 long-term memory, 37
 means–end strategy, 44
 Missing Dollar Problem, 38–39
 multidimensional scaling, 45
 pattern recognition, 36
 power of metaphors, 3
 search spaces, 42–44
 sensory store, 36
 short-term memory, 36
 summary, 46
 top–down processing, 37
 up–down metaphor, 36
 Venn diagram, 39
Spatial visualization, 1, 84, 192
Split-attention effect, 115, 117
StarLogo, 151
Static pictures, 79–82, 173
STM, *see* Short-term memory
Studies in Reflecting Abstraction, 132

Subject Index

Symbol(s)
  amodal, 95, 96
  ANIMATE, 169
  intention, 4
  perceptual, 6–10
  perceptual symbols hypothesis, 173
  systems, amodal, 6, 7
  thinking with, 3

T

Temperature Problem, 30
Temporal contiguity principle, 118
ThinkerTools software, 149, 150
Top–down processing, 37

U

Up–down metaphor, 36
U.S. Department of Education, recommendations, 175
User Illusion, The, 187

V

Venn diagrams, 39, 93, 95, 173
Verbal information, phonological loop for rehearsing, 113
Verbal memory, images and, 49
Verbal overshadowing effect, 54
Verbal thinking, 2
Viewing of pictures, 75–86, 187
  animated pictures, 82–84
  Companion Cognitive Systems, 79, 81
  decorative pictures, 76–78
  evocative pictures, 75–76
  finding slope of slide, 80
  political cartoons, 76
  proportional reasoning, 84
  Radiation Problem, 82
  spatial visualization, 84
  static pictures, 79–82, 84
  static versus animated pictures, 84–86
  summary, 86
  tank problems, 85
  topic interest, model of, 78
  Watergate, 76
Virtual reality, 135–145, 190
  desensitizing procedure, 136
  ecological psychology and multimedia, 137–139
  examples, 135–136
  exploring environments, 136–137
  information-processing models, 136, 138
  negotiation skills, training of, 142
  perceptual cycle, 138
  Pong, 135
  recognition-primed model, 141
  schemata, 137
  SimCity, 135
  situation awareness, 139–141
  summary, 144–145
  theory of perception, 137
  virtual biking, 143–144
  virtual driving, 144
  virtual military training, 141–142
Visions 2020.2, 176
Visual Display of Quantitative Information, The, 99, 166, 188
Visualization
  action and, 11
  amodal symbols and, 96
  computer models, 150
  power of in scientific discovery, 1
  school curricula supporting, 171
  spatial, 1, 84, 192
  tool, eChem, 156, 191
Visual Mathematics, 105
Visual memory, 13–14
Visual metaphor, making abstract concrete, 2
Visual simulation hypothesis, 9
Visual thinking
  examples of, 1
  importance of, 11
  words reflecting power of, 1

W

Where Mathematics Comes From: How the Embodied Mind Brings Mathematics into Being, 38, 185
Word(s), *see also* Images versus words; Words and pictures
  classifying into sentence, 50
  high-imagery, 48
  vertical alignment of, 10
Word Association Task, 9
Words and pictures, 113–123, 189
  bridging inference, 122
  cognitive load theory, 115–117
  cognitive overload, 115, 116
  coherence principle, 118
  episodic buffer, 115
  expertise reversal effect, 117
  farm toys, 121
  Galloping Horse Problem, 122
  individual differences principle, 118
  inferences, 122–123
  integration in working memory, 113–115
  Mayer's multimedia theory, 117–119

modality principle, 118
modified instructions physically integrating picture and words, 116
multimedia instruction, design principles, 118
multimedia model, 117
multimedia principle, 118
predictive inference, 122
redundancy effect, 116, 118, 119
simulated actions, 120–122
spatial contiguity principle, 118
split-attention effect, 115, 117
summary, 123
temporal contiguity principle, 118
verbal information, phonological loop for rehearsing, 113
words and images, 119–120
working memory, model of, 114
Working memory
   cognitive demands on, 130
   model of, 114